Stem Anatomy of *Dalbergia* and *Diospyros* Species from Madagascar

Bako Harisoa Ravaomanalina • Alan Crivellaro
Fritz Hans Schweingruber

Stem Anatomy of *Dalbergia* and *Diospyros* Species from Madagascar

with a Special Focus on Wood Identification

Springer

Bako Harisoa Ravaomanalina
Department of Plant Biology and Ecology
University of Antananarivo
Antananarivo, Madagascar

Alan Crivellaro
Department of TESAF
University of Padua
Legnaro, Padova, Italy

Fritz Hans Schweingruber
Institute of Forest, Snow and Landscape Research – WSL
Birmensdorf, Switzerland

ISBN 978-3-319-51145-0 ISBN 978-3-319-51147-4 (eBook)
DOI 10.1007/978-3-319-51147-4

Library of Congress Control Number: 2017936753

Printed on acid-free paper

This Springer imprint is published by Springer Nature
The registered company is Springer International Publishing AG
The registered company address is: Gewerbestrasse 11, 6330 Cham, Switzerland

Bako Harisoa RAVAOMANALINA is botanist and lecturer at the Department of Plant Biology and Ecology, University of Antananarivo in Madagascar. She has a PhD in Ecology. She is member of the CITES scientific authority and has experiences on Malagasy species listed in the Appendices of CITES (Convention on International Trade in Endangered Species of Wild Fauna and Flora). Her work is focusing on wood anatomy.
For this book, she collected the plant material, prepared the slides, realized the anatomical descriptions and photomicrographs, and managed the graphic design of the atlas.

Alan Crivellaro has degree in Wood Science, a MS in Forestry and a PhD in Ecology from the University of Padova, Italy.
Currently he is assistant professor in wood science at Dept. TESAF - University of Padova. His primary research interests have been wood, bark and pith anatomy to investigate the hydraulic, mechanical, and/or other roles of different stem tissues.
For this book, he contributed to the anatomical descriptions, and to the layout of the atlas.

Fritz Hans Schweingruber has a Ph.D. in Botany (Bern University, Switzerland). He was head of the research group Tree Ring and Site at the Swiss Federal Institute of Forest, Snow and Landscape Research (WSL) and professor at the Institute of Botany at the University of Basel, Switzerland. His work is documented by many scientific papers and books on dendrochronology and wood anatomy.
He supervised the material preparation, anatomical descriptions, and atlas design of the present book.

Acknowledgements

This research project is a result of collaboration between the Department of Plant Biology and Ecology (DBEV), University of Antanarivo, Madagascar, and The Swiss Federal Research Institute WSL, Switzerland. It was funded by a grant from the Swiss Government Excellence Scholarshio (No. 2013.0888/Madagascar/OP). Fieldwork expeditions were financially supported by a projet from the International Tropical Timber Organization (ITTO).

First of all, I wish to thank my husband Jean Julot Andrianarivoson, my son Diary and my daughter Emily.

I have to thank Pd. Dr. Ulf Büntgen, head of the dendroecology group, Swiss Federal Research Institute WSL, who accepted me in his group and showed enthusiasm and interest in my work.

This work was inspired by the results of my long work as Permanent Secretary within the CITES Scientific Authority at the Department of Plant Biology and Ecology, set up by the listing of Malagasy *Dalbergia* and *Diospyros* in Appendix II.

I would particularly like to thank Pete Lowry II and Georges Schatz (Missouri Botanical Garden, Missouri USA), without their taxonomic support, this study would not have been possible.

I am grateful to the staff of Missouri Botanical Garden in Madagascar and in Paris for their technical support. More specific thanks to Richard Randrianaivo, Richardson Razakamalala, Bernard Roger, Charles Rakotovao, Rakotonirina Benja, Fenonirina Rakotoarison, Felana Rakoto, Emeline, Ninah Sandratriniaina, Natacha Arimalala and Lucia Ravelomalala for collecting samples in the field.

I thank also the Madagascar National Parks and the "Direction Générale des Forêts" for providing the research and collect permit.

I am grateful for useful comments and suggestions of Prof. Pieter Baas (Netherlands Centre for Biodiversity Naturalis, Leiden), Prof. Jorgho Richter (Thünen-Institut, Hamburg) and Dr. Hans Beeckman (Royal Museum for Central Africa, Brussels). I want to express special thanks to Dr. Ruth Landolt (WSL) for substantial language corrections.

I want to thank the Swiss Federal Research Institute WSL which provided me a scientific infrastructure at Birmensdorf. Special thanks to Fano Rajaonary, Lena Hellmann, Lenka Matějů and Volodymyr Trotsiuk for preparing the plant distribution maps.

Many persons have invested their precious time and helped us in making this book, we are very grateful for their motivating cooperation. Thus special thanks go to ALL my co-workers in DBEV and WSL for the many constructive comments on our work and also just for the many pleasant moments we spent together.

The authors would like to thank the CITES management authority of Switzerland for financial support.

Bako Harisoa Ravaomanalina

Table of Contents

1. Introduction

Background

Madagascar has an exceptional biodiversity which is characterized by a high level of endemisms with a remarkable variety of bioclimates and habitats, including humid forests, dry deciduous forests, dry thicket forests, littoral forests, coastal zones and central high mountains. Malagasy natural forests are renowned for precious woods, which are a main forest product with a considerable added value on both international and national markets. The precious woods of Madagascar are mainly produced by members of 2 genera, *Dalbergia* and *Diospyros*. A total 48 species of *Dalbergia* are recognized, only one of which is non endemic (Bosser & Rabevohitra 2005), and some of which produce rosewood and palisander timber.

The taxonomy of *Diospyros* is currently being revised. A total of 85 described species are currently recognized, 82 of which endemic and 3 are not endemic, as summarized in the catalogue of Vascular Plant of Madagascar (www.tropicos.org/project/mada), and at least 130 additional species remain to be named and de scribed (Pete Lowry, pers. comm.)

These timbers have a significant economic role in Madagascar. Villagers use *Dalbergia* species as fuel. Rosewood and palisander species are exported as raw logs or finished products as furniture and handicrafts. Ebony wood (*Diospyros* spp.) is used for the manufacture of luxury goods, cutlery, brushes, marquetry, canes, lutes, wind musical instruments and piano keys. This black wood is highly demanded on the international market because of its colour, sturdy, hardness and resistance.

The continuing increase of international demand encourages illegal trade and harvesting of these species. Selective and excessive harvesting of mature, seed-bearing specimens leads to rarefaction of large-diameter trees. Extensive harvesting initiated small-diameter logging of trees for the handcraft products and musical instruments. Wood constitutes nearly the total energy supply especially in the region of Morondava.

Over 90 % of exported products are logs and sawn wood and only 10 % of wood are sold on local markets.

Timber identification on the international market is not possible from logs because morphological traits used in species identification such as flowers, bark and fruits have been removed. Therefore, it is possible for customs authorities to distinguish legally from illegally harvested logs. For these reasons, Malagasy precious woods *Dalbergia* and *Diospyros* have received protection under CITES Appendix II in 2013. The listing of these species helps to ensure the legality of trade, the traceability of wood and wood products, and it should help to ensure that the exploitation is not detrimental to the survival of these species. Identification is the basic element of CITES enforcement. To facilitate adequate implementation of the Appendix II listing of *Dalbergia* spp. and *Diospyros* spp., Madagascar presented an action plan in which establishment of reliable and fast identification techniques for logs and wood products need to be developed and implemented.

Considerable effort occurred to enhance anatomical, chemical and physical methods to address this problem in all countries with natural occurrences of these species. In general, wood anatomy is the most used and appropriate method to identify timber species. Gasson *et al.* (2010) pursued quantitative anatomical characters using PCA and Bayes comparisons, and managed to identify all *Dalbergia nigra* samples. Miller and Wiemann (2006) compared the wood of *Dalbergia nigra* and *Dalbergia spruceana* and used the different fluorescence of water extracts and wood density to separate the two. Kite *et al.* (2010) found the flavonoid that is unique to *Dalbergia nigra*, at least in the 15 American, African and Asian species

B.H. Ravaomanalina et al., *Stem Anatomy of Dalbergia and Diospyros Species from Madagascar*, DOI 10.1007/978-3-319-51147-4_1

compared. Pigozzo *et al.* (2010) have suggested to distinguish *Dalbergia* species, especially *Dalbergia nigra* and *Dalbergia spruceana* by using infrared spectrometry. A rapid chemical method of analysis was developed by Lancaster and Espinoza (2012) to differentiate commonly traded species, among them *Dalbergia baronii* and *Dalbergia madagascariensis*. Mass spectrometry was applied to distinguish different species of *Dalbergia* at the international level, however species from Madagascar need more detailed analyses (Espinoza et al. 2015; McClure et al. 2015; Lancaster & Spinoza, 2012). However, except the wood anatomical descriptions recently given by P. Détienne, available on InsideWood (Wheeler, 2011), extensive studies of Malagasy *Dalbergia* species in relation to wood identification are still lacking. Anatomical descriptions of *Diospyros lotus* and *Diospyros kaki* have been made by Schweingruber (1990) and Saldari *et al.* (2008). Changes in ray arrangement of *Diospyros lotus* was analyzed by Myśkow and Zagórska-Marek (2012) and some physical, biometry and mechanical strength properties were determined by Kiaei and Bakhshi (2014). Anatomical descriptions of some Malagasy *Diospyros* species are available on the wood database InsideWood. Moreover, not all the descriptions are based on vouchered wood samples, and we have to take in mind that some woods may not have been correctly identified (IAWA, 2011) in addition to that, the genus *Diospyros* is currently on taxonomic revision. All descriptions studied in this book are based on vouchered wood samples and they are available in collections. The species identification was made by the taxonomy specialists Georges Schatz and Pete Lowry II, Missouri Botanical Garden, France and US.

About the manual

This manual replies to the expectations of the action plan for *Diospyros* and *Dalbergia* about the establishment of a reference collection and identification system for CITES-listed species. Since this is the first time that an anatomical atlas about Malagasy precious wood was planned we decided to give an overview about xylem and bark from twigs, branches and stems. Special focus was given on stem wood identification. This study will address a limited number of species, relevant to the international trade. The monographic presentation are arranged by genera, and within them by species in alphabetical order.

A key for the identification of *Dalbergia* and *Diospyros* species is provided on the base of stem wood anatomical features.

This atlas covers 19 *Dalbergia* species and 31 *Diospyros* species. Among the described species, *Dalbergia bracteolata* and *Diospyros ferrea* are the only two species which are not endemic to Madagascar. Ten *Diospyros* species were not described and published taxonomically (*Diospyros acutiflora, D. antongiliensis, D. brevipedicellata, D. chitinophora, D. maculata, D. maxima, D. olacinoides, D quadrangularis, D. rubripetiolata, D. tropophylloides*). The codification of anatomical features is based on one individual, represented by the stem, branch and twig. Since *Dalbergia* and *Diospyros* species grow in different vegetation formations in Madagascar, ecological adaptations are expressed by various anatomical structures.

2. Materials and Methods

Origin of the material studied

The sampling of plant material was carried out in Madagascar in July 2013 and January 2014. The sampling sites are indicated on the bioclimatic map. All the species were collected in their natural habitat. Species identification was carried out by the taxonomists of Missouri Botanical Garden both in Madagascar and in Missouri. General botanical information were carried out by Missouri Botanical Garden Madagascar (2009, 2014) and the Department of Plant Biology and Ecology (2010, 2011). Data distributions were derived from the Catalogue of Vascular Plant of Madagascar (www.tropicos.org). Consequently, database per specimens used in the atlas are available on the same web site. Herbarium specimens of the presented material can be consulted at the international herbaria Missouri (MBG), Paris (MNHN), Brussels (RMCA) and Zürich (Botanical Garden) while wood specimens were available from wood collection of Tervuren in Brussels (RMCA).
The sampled species are summarized on the Table 1a and 1b. The number of the wood samples, deposited in the Tervuren Wood collection of the Royal Museum for Central Africa (Belgium) is indicated with Tw. Slides collection are also available at WSL Birmensdorf and at the Department of Plant Biology and Ecology in Antananarivo (Madagascar).
Ten *Diospyros* species on bold showed in Table 1b don't have the authors name because they are already identified by the specialist but not yet officially published (Schatz and Lowry comm. pers.).
Botanical descriptions and the use of plants are briefly given per species. Local names in English (EN), in French (FR) and in Malagasy (ML) are provided. Ecological conditions of the natural habitat as elevation, vegetation and bioclimate are extremely useful for the identification. The presence of the species on the defined previously protected areas (PA) and the new one (NPA) contributes to the protection of these precious woods. The origin of the sampled species is always mentioned on the distribution map.

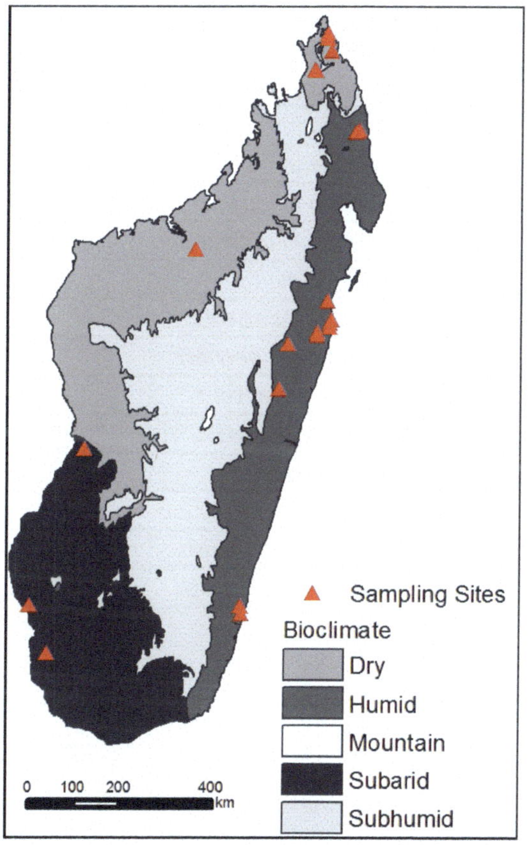

Sampling sites of Malagasy *Dalbergia* and *Diospyros* species.

B.H. Ravaomanalina et al., *Stem Anatomy of Dalbergia and Diospyros Species from Madagascar*, DOI 10.1007/978-3-319-51147-4_2

Table 1a. Descriptive information on *Dalbergia* samples analysed.

1) Collector (CR: Charles Rakotovao; RAF: Rakotoarison Fenonirina; RBE: Roger Bernard; RIR: Richard Randrianaivo; RZK: Richardson Razakamalala; T: Department of Plant Biology and Ecology, University of Antananarivo); 2) Sample identification number; 3) Scientific name; asterisks indicate species not endemic to Madagascar; 4) Author name; 5) Locality of sampling; 6) Forest type from which the plant was sampled (D.: Deciduous); 7) Identification number of the specimens available at the Tervuren Xylarium, TW no.

1) Collector	2) Sample	3) Species	4) Author	5) Locality	6) Forest type	7) Tw no.
T	12	*Dalbergia baronii*	Baker	Didy	Humid	66656
RZK	7709	*Dalbergia bathiei*	R.Vig.	Tampolo	Humid	66551
CR	6513	**Dalbergia bracteolata*	Baker	Kirindy	Dry, D.	66454
RZK	7712	*Dalbergia chapelieri*	Baill.	Tampolo	Humid	66553
T	24	*Dalbergia emirnensis*	Benth.	Mt Français	Dry, D.	66657
RIR	2416	*Dalbergia greveana*	Baill.	Ankarana	Dry, D.	66477
RIR	2430	*Dalbergia lemurica*	Bosser & R.Rabev.	Oronjia/Rigny	Dry, D.	66490
RBE	2247	*Dalbergia louvelii*	R. Vig.	Betampona	Humid	55506
RAF	13	*Dalbergia madagascariensis*	Vatke	Makirovana	Humid	55566
RZK	7704	*Dalbergia maritima*	R. Vig.	Betampona	Humid	66546
RIR	2478	*Dalbergia mollis*	Bosser & R.Rabev.	Ankarafantsika	Dry, D.	66530
T	37	*Dalbergia monticola*	Bosser & R.Rabev.	Didy	Humid	66658
CR	6512	*Dalbergia neoperrieri*	Bosser & R.Rabev.	Kirindy	Dry, D.	66453
RZK	7699	*Dalbergia orientalis*	Bosser & R.Rabev.	Betampona	Humid	66541
RIR	2440	*Dalbergia peltieri*	Bosser & R.Rabev.	Oronjia/Rigny	Dry, D.	66499
T	25	*Dalbergia pervillei*	Vatke	Mt Français	Dry, D.	66659
RIR	2418	*Dalbergia purpurascens*	Baill.	Ankarana	Dry, D.	66479
RIR	2446	*Dalbergia suaresensis*	Baill.	Mt Français	Dry, D.	66505
RIR	2470	*Dalbergia trichocarpa*	Baker	Ankarafantsika	Dry, D.	66551

Table 1b. Descriptive information on *Diospyros* samples analysed.

1) Collector (CR: Charles Rakotovao; EME: Emeline; RAF: Rakotoarison Fenonirina; RBE: Roger Bernard; RIR: Richard Randrianaivo; RNA; Razafimamonjy Nivo Alisoa; RZK: Richardson Razakamalala; T: Department of Plant Biology and Ecology, University of Antananarivo); 2) Sample identification number; 3) Scientific name; asterisk indicate species not endemic to Madagascar; 4) Author name (ined.: inedited); 5) Locality of sampling; 6) Forest type from which the plant was sampled (D.: Deciduous, T.: Thicket); 7) Identification number of the specimens available at the Tervuren Xylarium, TW no.

1) Collector	2) Sample	3) Species	4) Authors	5) Locality	6) Forest type	7) Tw no.
T	33	*Diospyros aculeata*	H.Perrier	Beheloka	Dry, T.	66660
RIR	7724	*Diospyros acutiflora*	ined.	Andranotsara	Humid	66562
T	27	*Diospyros analamerensis*	H.Perrier	Mt Français	Dry, T.	66661
RIR	2405	*Diospyros ankifiensis*	H.Perrier	Ankarana	Dry, D.	66468
RAF	6	*Diospyros antongiliensis*	ined.	Makirovana	Humid	66517
RIR	2461	*Diospyros bernieriana*	(Baill.) H.Perrier	Sahafary	Dry, D.	66517
RBE	2248	*Diospyros brevipedicellata*	ined.	Betampona	Humid	66368
RBE	2264	*Diospyros calophylla*	Hiern	Analalava	Humid	66384
RIR	2403	*Diospyros chitinophora*	ined.	Ankarana	Dry, D.	66466
RZK	7714	**Diospyros ferrea*	(Willd.) Bakh.	Andranotsara	Humid	66555
RZK	7703	*Diospyros fuscovelutina*	H.Perrier	Betampona	Humid	66545
CR	6516	*Diospyros humbertiana*	H.Perrier	Kirindy	Dry, D.	66457
RNA	1	*Diospyros latispathulata*	H.Perrier	Ranobe	Dry, T.	66533
RZK	7700	*Diospyros leucocalyx*	Hiern	Betampona	Humid	66542
RBE	2240	*Diospyros maculata*	ined.	Betampona	Humid	66360
T	42	*Diospyros mangabensis*	Aug.DC.	Andasibe	Humid	66662
RBE	2239	*Diospyros maxima*	ined.	Betampona	Humid	66359
EME	21	*Diospyros microrhombus*	Hiern	Mahabo Mananivo	Humid	66408
RIR	2411	*Diospyros occlusa*	H.Perrier	Ankarana	Dry, D.	66452
RIR	2439	*Diospyros olacinoides*	ined.	Oranjia/Rigny	Dry, D.	66498
T		*Diospyros parifolia*	(Schltr.) Bakh.	Didy	Humid	66663
RIR	2452	*Diospyros perrieri*	Jum.	Mt Français	Dry, D.	66511
EME	20	*Diospyros pervilleana*	(Baill.) G.E.Schatz & Lowry	Mahabo Mananivo	Humid	66407
CR	6504	*Diospyros platycalyx*	Hiern	Kirindy	Dry, D.	66445
RZK	7710	*Diospyros quadrangularis*	ined.	Tampolo	Humid	66552
RZK	7715	*Diospyros rubripetiolata*	ined.	Andranotsara	Humid	66556
RBE	2253	*Diospyros sclerophylla*	H. Perrier	Tampolo	Humid	66373
RZK	7713	*Diospyros squamosa*	Bojer ex A.DC.	Andranotsara	Humid	66554
T	32	*Diospyros tropophylloides*	ined.	Mt Français	Dry, D.	66664
EME	4	*Diospyros velutipes*	(H. Perrier) G.E.Schatz & Lowry	Mahabo Mananivo	Humid	66394
RIR	2407	*Diospyros vescoi*	Hiern	Ankarana	Dry, D.	66470

Plant material preparation

The individual chosen in its natural habitat appeared normal and healthy. A portion of the main stem was cut near the plant's basis. Twig and branch samples were also collected from the same individual. Branch samples were collected from the first branch with the height dependent on the position of the branch and ranging from 5 to 20 m.
Plant material was stored in a sealed plastic bag to which several drops of 40 % ethanol were added. Each sample was labelled with a permanent marker pen containing the collector codes and the identification number of the sample(s).
In the laboratory, from each branch and stem sample, a 1 cm^3 sample was split from the outermost wood, including the bark. For twig samples, special care was taken to preserve the bark and the pith on the same sample. The thin sections (15 - 25 µm) have been made using disposable blades and GLS microtome (Gärtner and Scweingruber, 2013). The section were stained with Astra blue and Safranin O, dehydrated with alcohol and xylene, and mounted in Canada balsam (Chaffey 2002, Gärtner and Schweingruber, 2013). Sections with bark were initially bleached in Eau de Javel for 10 to 15 minutes before staining.
A transmission-light microscope was used to observe slides while polarized filter were applied for observing crystal presence and cell wall thickness. Sections were microscopically inspected using magnifications of 20-1000 times and photographed by using a digital camera mounted on the microscope (Olympus BX41). Magnifications are indicated in µm (0,001 mm) above a black scale bar in each picture.
For all species the anatomy of pith, primary xylem, branch xylem, twig and stem bark and stem xylem were described. Differences between branch and stem xylem were determined.

3. Definition of Anatomical Features

Xylem anatomy is described on the basis of the IAWA List of Microscopic Features for Hardwood Identification (Wheeler *et al.* 1989). However, three relevant anatomical features for *Dalbergia* and *Diospyros* species identification have been added to the IAWA list. For those additional features, we used the code in the form "X.#" where X is the modified relative IAWA character number and "#" is a consecutive number.

9.1 *- Vessels solitary and in short radial multiples of 2 to 3 elements simultaneously present*

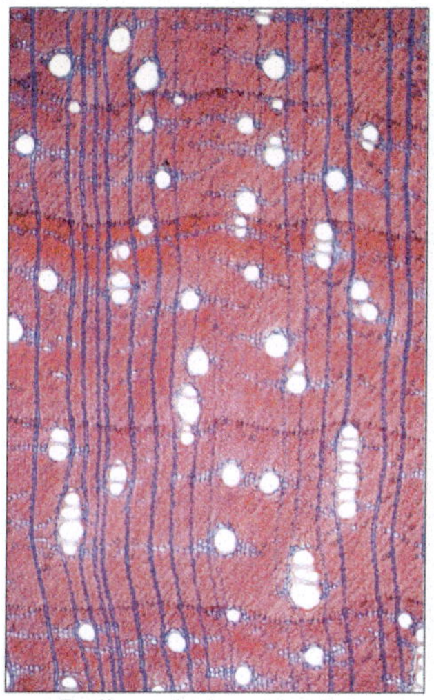

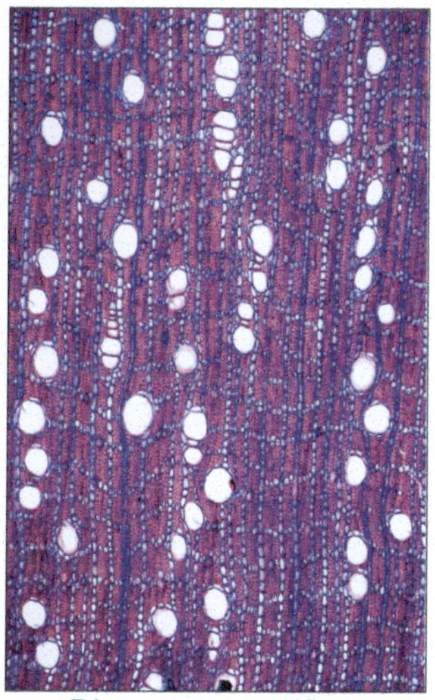

Dalbergia bracteolata *Diospyros acutiflora*

96.1 *- Rays uni- and biseriate simultaneously present*

Ray width is determined on tangential sections by counting the number of cells in the widest part of the rays, perpendicular to the ray axis. Rays uni- and biseriate are often simultaneously presents in both *Dalbergia* and *Diospyros*, in this case width predominance is always determined.

102.1 *- Ray height < 100 µm*
102.2 *- Ray height 100-200 µm*
102.3 *- Ray height 200-500 µm*
102.4 *- Ray height > 500 µm*

Detailed ray height data are given in descriptions because this feature may be helpful in distinguishing between species. Ray height is quite variable in *Diospyros* with markedly heterogeneous rays (IAWA, 1989). Ray height is not exceeding 1 mm in both *Dalbergia* and *Diospyros*. Four classes are determined for the total ray height in tangential section, along the ray axis. Size measurements of all wood elements are presented as minimum, mean and maximum (e.g. 10-24-50 µm).

B.H. Ravaomanalina et al., *Stem Anatomy of Dalbergia and Diospyros Species from Madagascar*, DOI 10.1007/978-3-319-51147-4_3

4. Identification Keys

How to use the keys

The present keys focus on stem wood identification, in most cases referred as sapwood, but sometimes from the transition between sapwood and heartwood.
Within the genus *Diospyros*, species are classified by ray width (exclusively uniseriate rays, uni- and biseriate rays and rays 2 to 3 cells wide). Most of the species present uni- and biseriate rays at the same time. Then species are categorized according to vessels grouping. In *Dalbergia*, species are grouped by porosity, then by ray type and subsequently by ray width. Correct identification can only be achieved through comparison with illustrations and descriptions presented in the main section of the atlas. Two separate species which fall in the same category additional features such as width of banded parenchyma and number of storied structure can be useful to reach the species level.

B.H. Ravaomanalina et al., *Stem Anatomy of Dalbergia and Diospyros Species from Madagascar*, DOI 10.1007/978-3-319-51147-4_4

Stem wood identification key of Dalbergia

Ring porosity	Ray type	Ray width	Species
Ring porous	Heterogeneous	2-3 cells	*Dalbergia purpurascens*
	Homogeneous	2-3 cells	*D. greveana* *D. peltieri*
Diffuse-porous	Heterogeneous	1 cell	*D. bracteolata* *D. maritima*
		1(2) cells	*D. bathiei*
		2-3 cells	*D. baronii* *D. monticola* *D. orientalis*
	Homogeneous	1 cell	*D. chapelieri* *D. madagascariensis*
		1(2) cells	*D. suaresensis* *D. trichocarpa*
		2-3 cells	*D. emirnensis*
	Heterogeneous and Homogeneous	1 cell	*D. pervillei*
		1(2) cells	*D. louvelii* *D. neoperrieri*
		2-3 cells	*D. lemurica* *D. mollis*

Stem wood identification key of Diospyros

Ray width	Vessels grouping	Species
1 cell	Vessels predominantly in short radial multiples	*D. aculeata* *D. acutiflora* *D. antongiliensis* *D. calophylla* *D. leucocalyx*
	Vessels solitary and in long radial multiples	*D. quadrangularis*
	Vessels in radial multiples	*D. analamerenis* *D. ankifiensis* *D. latispathulata* *D. maxima* *D. pervilleana*
1(2) cell(s)	Vessels predominantly in short radial multiples	*D. bernieriana* *D. chitinophora* *D. fuscovelutina* *D. humbertiana* *D. maculata* *D. microrhombus* *D. sclerophylla* *D. squamosa*
	Vessels in predominantly in long radial multiples	*D. brevipedicellata* *D. perrieri* *D. ferrea* *D. mangabensis* *D. occlusa* *D. olacinoides* *D. parifolia* *D. platycalyx* *D. tropophylloides* *D. velutipes* *D. vescoi*
2-3 cells	Vessels mostly solitary	*D. rubripetiolata*

5. *Dalbergia*

A total of 48 species of *Dalbergia* are recognized, some of them produce rosewood and palisander timbers.
Here we describe the stem anatomy of 19 *Dalbergia* species.

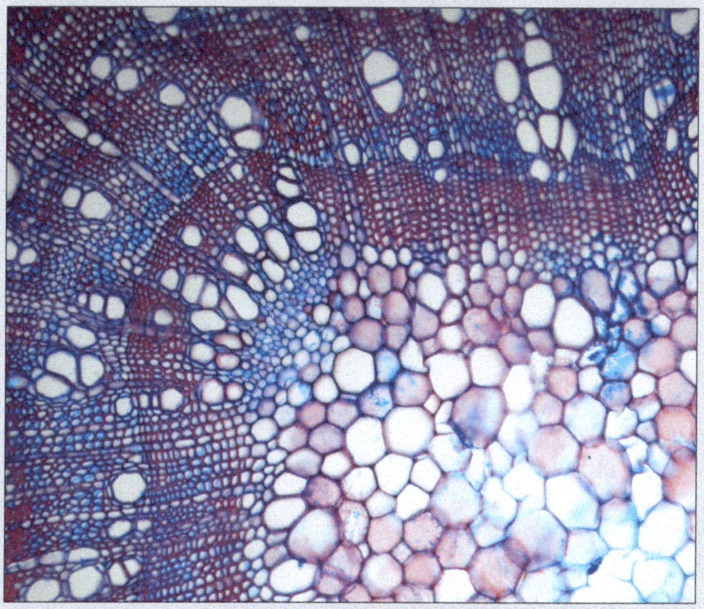

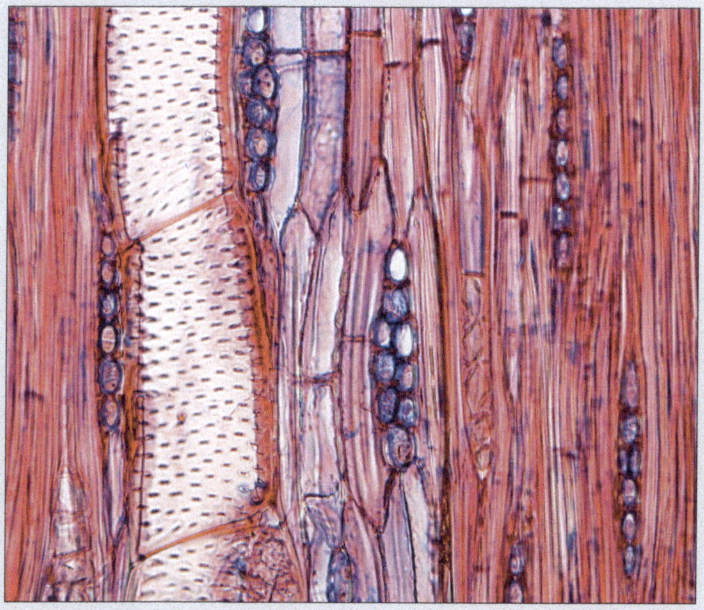

B.H. Ravaomanalina et al., *Stem Anatomy of Dalbergia and Diospyros Species from Madagascar*, DOI 10.1007/978-3-319-51147-4_5

Dalbergia baronii Baker

EN: Palisander FR: Palissandre ML: Voamboana fotsy

PLANT DESCRIPTION

Shrub to tree | 3-30 m high | 8-40 cm DBH
Endemic to Madagascar. Exploited species.

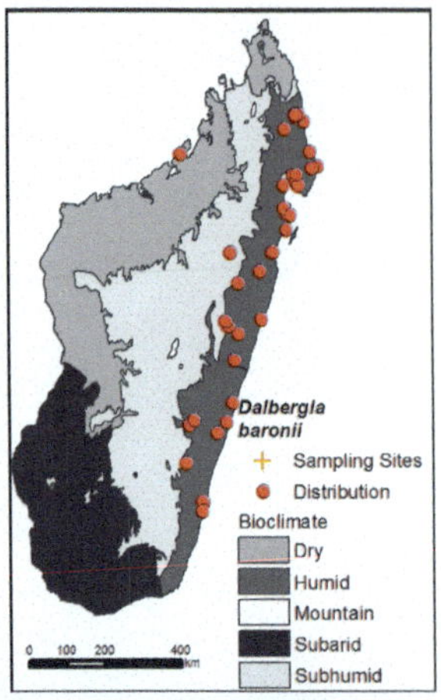

DISTRIBUTION

Elevation: 0-1000 m a.s.l.
Vegetation formation: forest, freshwater wetland, mangrove and saline wetland
Bioclimate: humid
Protected areas: Mananara Nord, Masoala and Ranomafana
New protected areas: Didy

Slides made from material collected in Didy.

PITH AND PRIMARY XYLEM ANATOMY

Pith: Shape of pith round. A few scattered thick-walled parenchyma cells present. Cells dimorphic. Simple pits in transverse cell walls. Pith not visible in polarized light.

Primary xylem: Vascular bundles not distinct. Vessels mostly solitary.

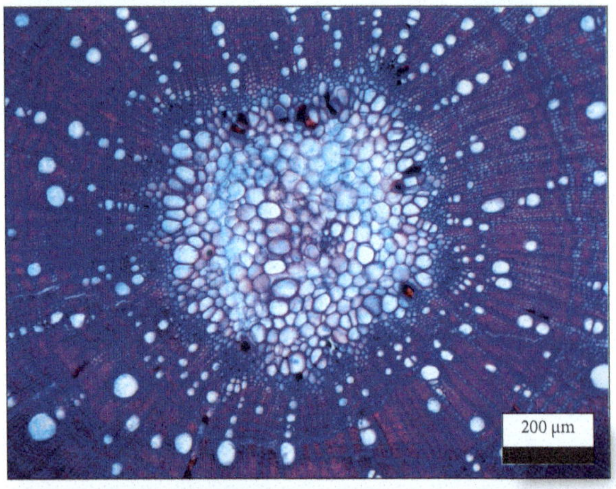

BRANCH XYLEM ANATOMY

Description: Growth rings boundaries indistinct or absent. Tangential diameter of vessels 50-89-130 µm, 7-10-17 vessels/mm^2. Vessel element length 120-163-190 µm. Intervessel pits small, 4-7 µm in diameter. Fibres thin- to thick-walled. Axial parenchyma in bands up to 3 and more than 3 cells. Rays predominantly uniseriate. Ray height 60-101-140 µm. Rays 7-9-12/mm.

Anatomical differences to stem xylem: Mean tangential diameter of vessel lumina in branches is slightly smaller (50-89-130/110-158-300 µm) while vessels number is slightly higher (7-10-17/4-7-10 vessels/mm^2). Ray height and number are slightly lower in the branch (60-101-250/50-114-150 µm; 7-9-12/11-12-14 rays/mm).

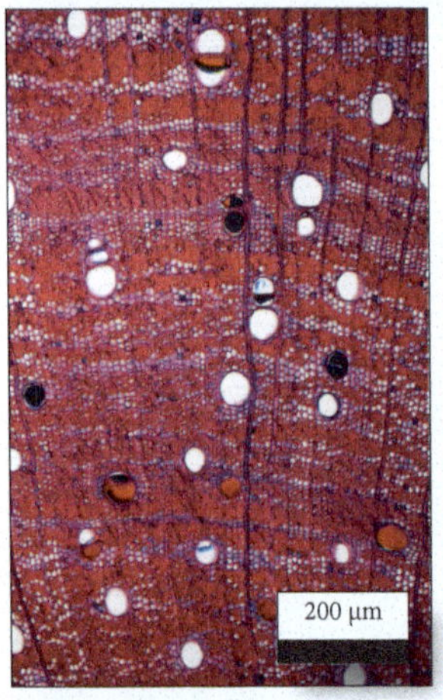

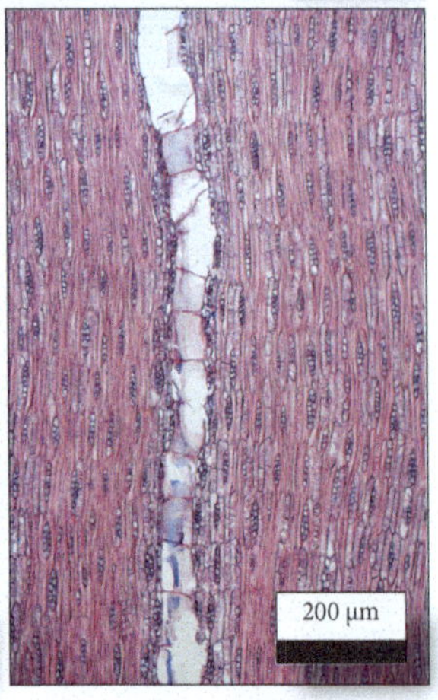

BARK ANATOMY

Branch bark: The phloem consists of alternating groups of fibres and collapsed and non-collapsed sieve tubes. Prismatic crystals present. The cortex consists of large and irregularly shaped cells and tangentially distributed fibres.

Stem bark: The phloem consists of alternating bands of fibres and collapsed and non-collapsed sieve tubes. Only some rays become dilated. Axial parenchyma dilatation present. Crystal sand present. Phellem homogeneous, it consists of regularly shaped rectangular cells. Lignified cells in phellem are distinct in polarized light.

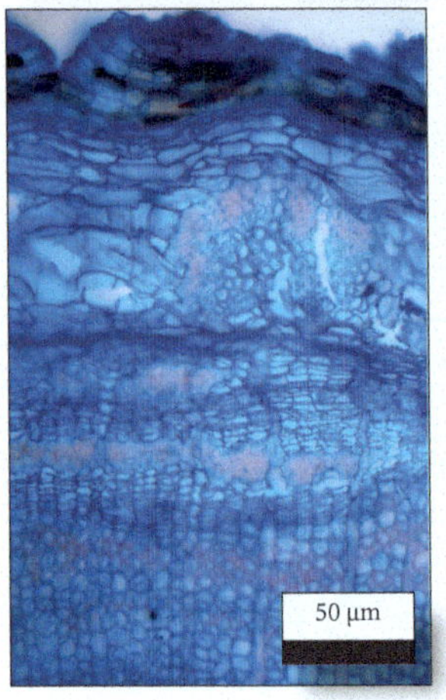

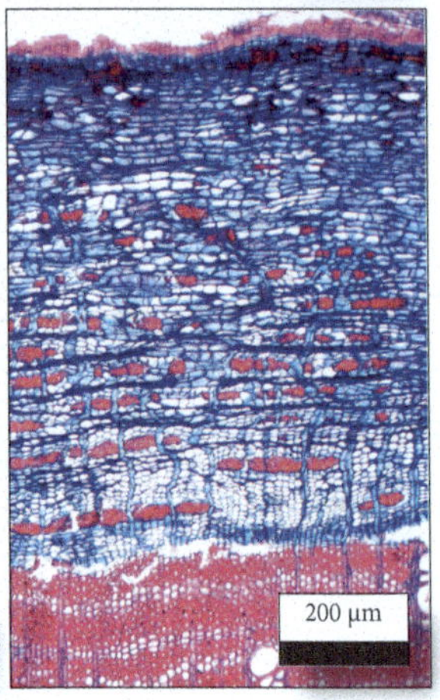

IAWA feature numbers
1 5 9 9.1 13 22 25 29 30 42 47 52 58 61 66 70 77 86 89 90 91 97 102.2 106 116 118 120 136 142 180 189 190 192 196

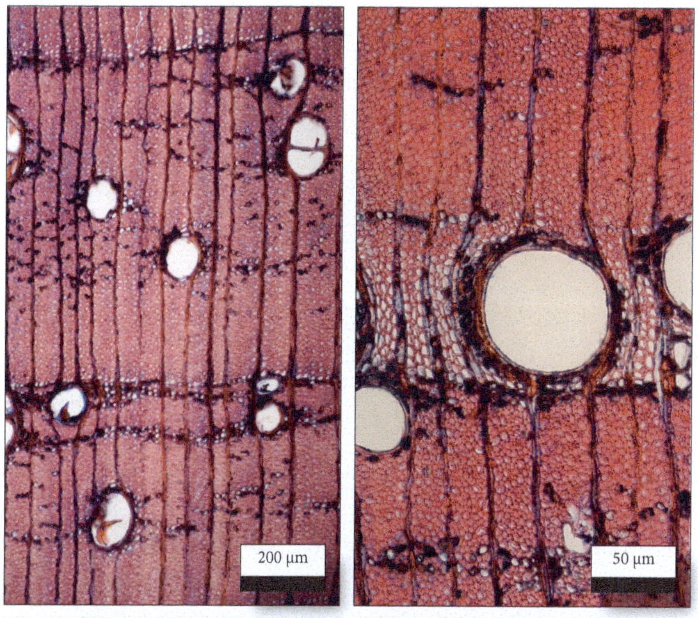

STEM XYLEM ANATOMY

Transverse section
Growth ring boundaries distinct.
Wood diffuse-porous.
Vessels solitary and in radial multiples of 2-5 elements.
Tangential diameter of the vessels 110-158-300 µm, 4-7-10 vessels/mm^2.
Gums and/or other deposits in heartwood vessels.
Fibres very thick-walled.
Axial parenchyma diffuse-in-aggregates and in narrow lines up to 3 cells. Seemingly parenchyma marginal bands present.
Rays 11-12-14/mm.

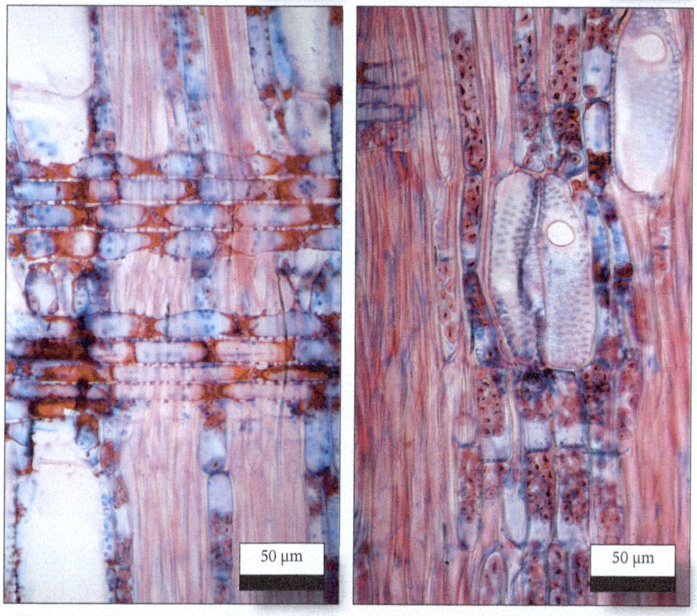

Radial section
Simple perforation plates.
Intervessel pits oval to circular and alternate, 4-7 µm in diameter.
Vestured pits present.
Vessel-ray pits with distinct borders; similar to intervessel pits in size and shape throughout the ray cell.
Vessel element length 130-190-210 µm.
Fibres with simple to minutely bordered pits.
Heterogeneous rays with body ray cells procumbent and one row of upright and square marginal cells.
Prismatic crystal in chambered axial parenchyma cells.

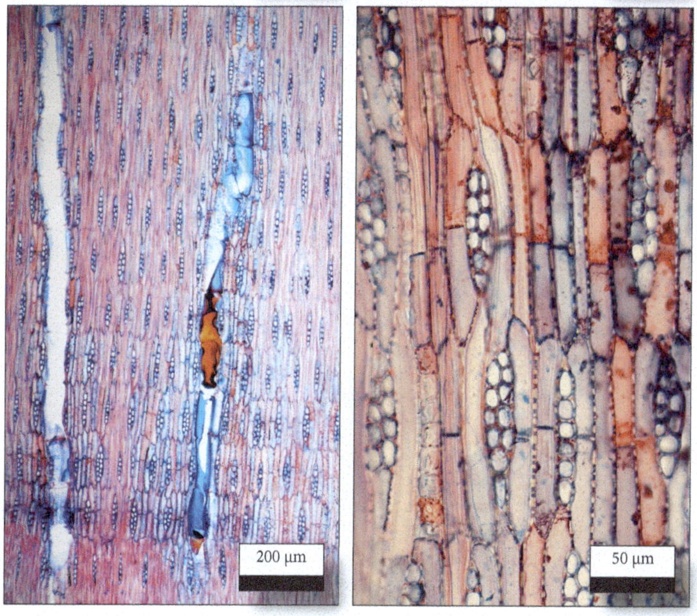

Tangential section
Rays predominantly biseriate.
Ray height 50-114-150 µm.
Fusiform parenchyma cells present. Two cells per parenchyma strand.
Rays, vessels and axial parenchyma irregularly storied. Number of ray tiers per axial mm: 13-14.

Dalbergia bathiei R. Vig.

EN: Palisander FR: Palissandre ML: Voamboana

PLANT DESCRIPTION

Shrub to tree | 3-30 m high | 10-22 cm DBH
Endemic to Madagascar. Exploited species.

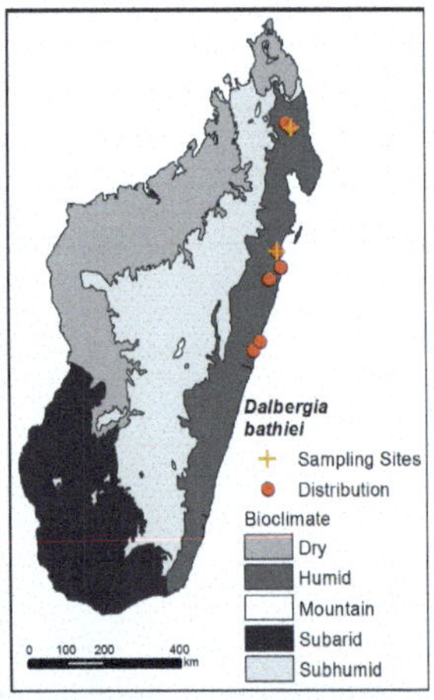

DISTRIBUTION

Elevation: 0-500 m a.s.l.
Vegetation formation: forest and freshwater wetland
Bioclimate: humid
Protected areas: Betampona and Marojejy
New protected areas: Tampolo

Slide made from material collected in Tampolo.

PITH AND PRIMARY XYLEM ANATOMY

Pith: Shape of pith triangular. Scattered thick-walled parenchyma cells present. Thin-walled and blue-stained cells present. Cells dimorphic. Dark staining substances in cell lumina of thick-walled cells. Prismatic crystals rare. Pits in transverse cell walls. Pith not visible in polarized light.

Primary xylem: Vascular bundles not distinct. Vessels solitary and in long radial multiples. Fibres thin- to thick-walled in radial rows.

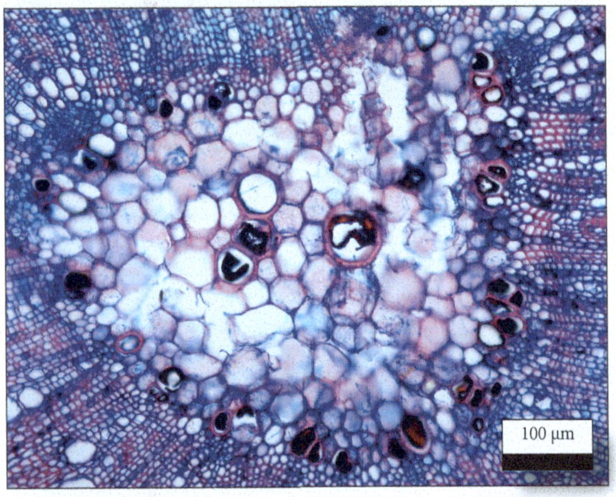

BRANCH XYLEM ANATOMY

Description: Growth rings boundaries indistinct or absent. Tangential diameter of vessels 80-102-130 µm, 8-9-12 vessels/mm^2. Vessel element length 50-94-200 µm. Intervessel pits small, 4-7 µm in diameter. Fibres thick- to very thick-walled. Axial parenchyma in bands 2-9 cells wide. Rays exclusively uniseriate. Ray height 50-104-150 µm, rays 7-10-14/mm.

Anatomical differences to stem xylem: Mean tangential diameter of vessel lumina and number are similar in the branch and in the stem (80-102-130/70-105-160 µm; 8-9-12/6-8-12 vessels/mm^2). Rays are slightly shorter in the branch (50-104-150/50-148-250 µm) while ray number is comparable (7-10-14/8-9-11 rays/mm).

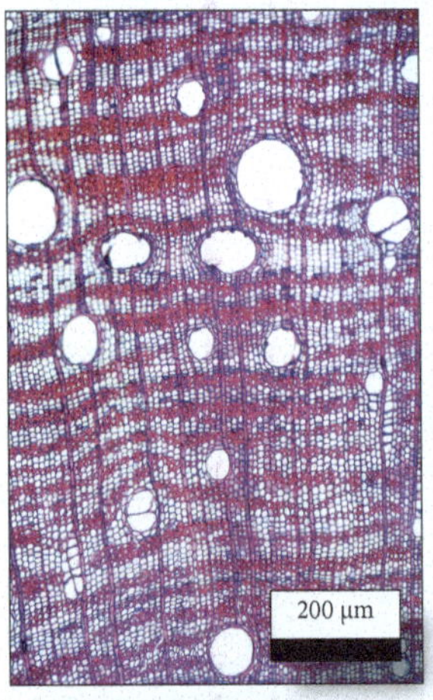

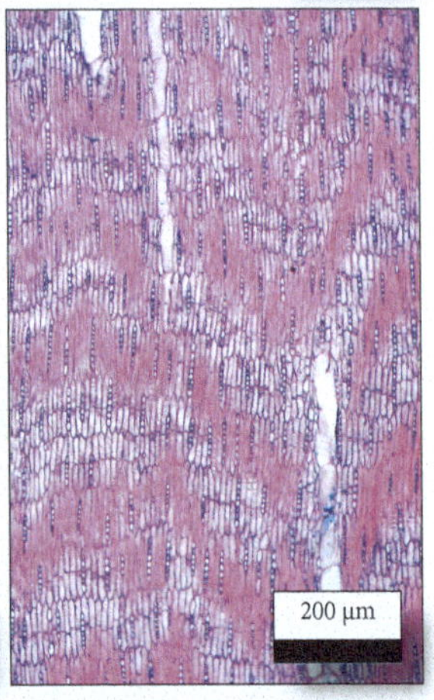

BARK ANATOMY

Branch bark: Sieve tubes arranged in tangential row. Collapsed sieve tubes present. The cortex consists of irregular blue-stained cells and groups of tangentially distributed fibres.

Stem bark: The phloem consists of alternating bands of fibres and collapsed and non-collapsed sieve tubes. Only some rays become dilated, axial parenchyma cell dilatation frequent. Crystal sand present. The phellem is homogeneous, it consists of homogeneous rectangular cells. Lignified cells in phellem are distinct in polarized light.

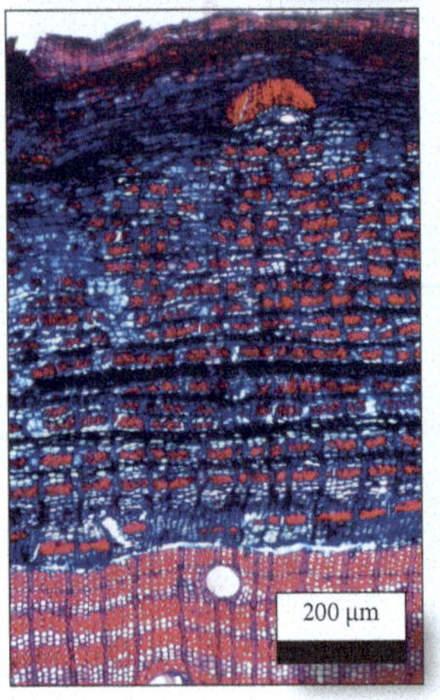

IAWA feature numbers

1 5 9 13 22 25 29 30 42 47 52 58 61 66 70 85 90 91 96.1 102.2 106 115 118 120 136 138 142 180 189 192 196

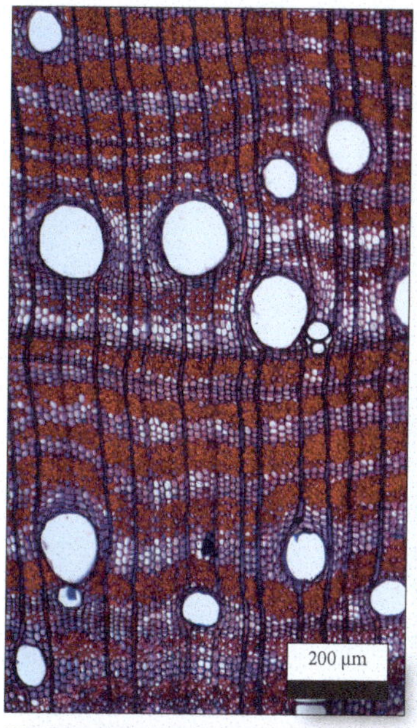

STEM XYLEM ANATOMY

Transverse section

Growth rings boundaries distinct.
Wood diffuse-porous.
Vessels mostly solitary. Tangential diameter of vessels 70-105-160 µm, 6-8-12 vessels/mm^2.
Gums present in heartwood vessel lumina.
Fibres very thick-walled.
Axial parenchyma in bands more than 3 (2-9) cells wide.
Rays 8-9-11/mm.

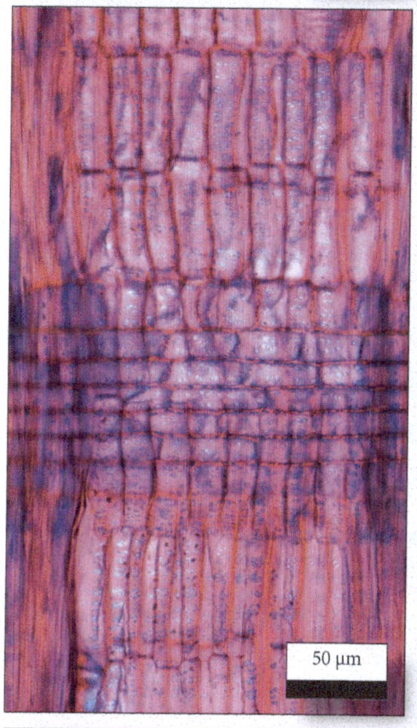

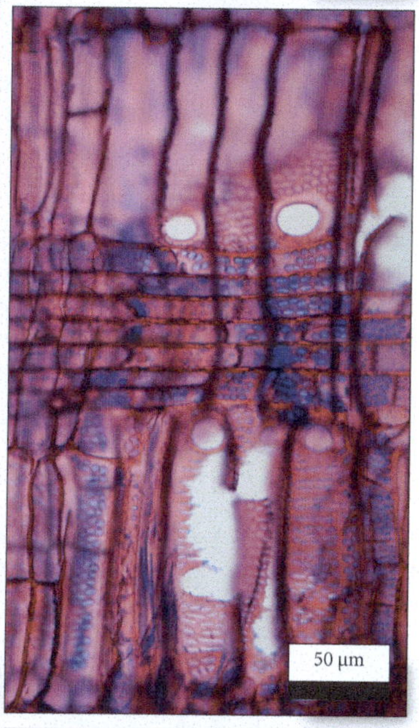

Radial section

Simple perforation plates.
Intervessel pits oval to circular and alternate, 4-7 µm in diameter.
Vestured pits present.
Vessel-ray pits with distinct borders; similar to intervessel pits in size and shape throughout the ray cell.
Vessel element length 210-239-270 µm.
Fibres with simple to minutely bordered pits.
Heterogeneous ray with body ray cells procumbent and one row of upright and square marginal cells.
One prismatic crystal per chamber in procumbent ray cells and in axial parenchyma cells.

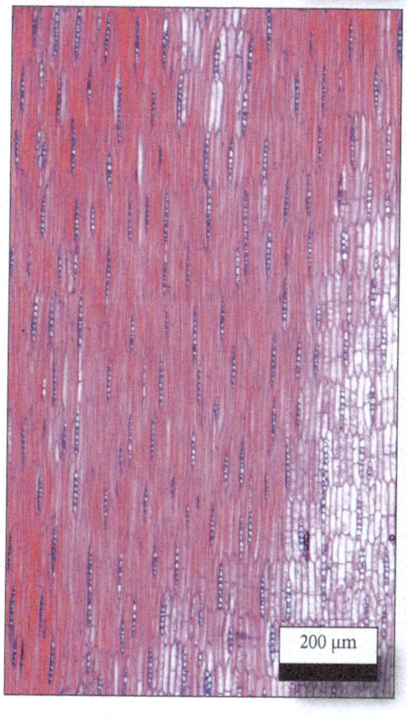

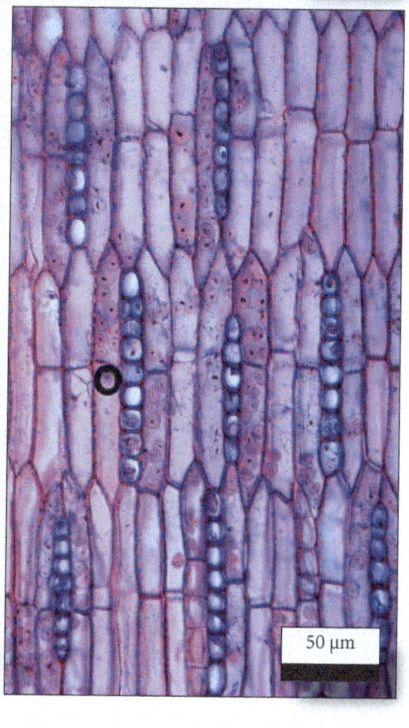

Tangential section

Rays predominantly uniseriate, a few bi-seriate rays present. Ray height 50-148-250 µm.
Fusiform parenchyma cells present. Two cells per parenchyma strand.
Rays and axial parenchyma irregularly storied.
Number of ray tiers per axial mm: 14-15.

Dalbergia bracteolata Baker

Synonymy: *D. grandidieri* Baill., *D. richardii* Baill., *D. goetzei* Harms.

EN: - FR: - ML: Felangoaka, Vahinta

PLANT DESCRIPTION

Woody liana | 2-7 m high | up to 7 cm DBH
Native to Madagascar, common in Africa mainland. Non exploited species.

DISTRIBUTION

Elevation: 0-1000 m a.s.l.
Vegetation formation: forest, grassland and woodland
Bioclimate: dry, sub-arid and sub-humid
Protected areas: Ankarafantsika, Ankarana, Bemaraha, Isalo, Kirindy-Mitea, Namoroka, Zombitsy-Vohibasia and Kirindy
New protected areas: Daraina complex, Makay massif, Mikea, Montagne des Français and Sahafary

Slides made from material collected in Kirindy-Menabe.

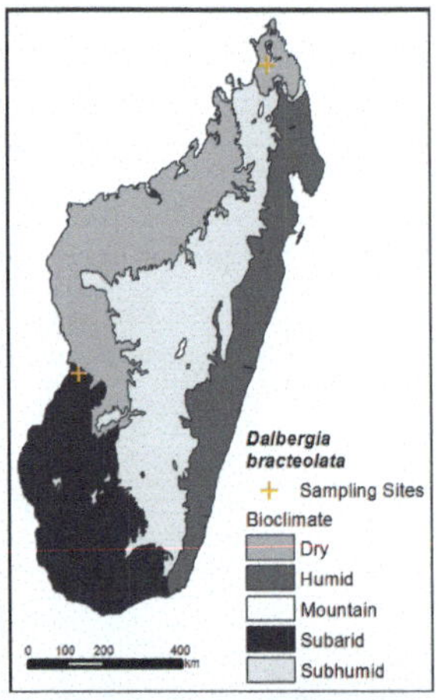

PITH AND PRIMARY XYLEM ANATOMY

Pith: Shape of pith oval. Thick-walled parenchyma cells present. Cells dimorphic. Simple pits in transverse cell walls. Pith not visible in polarized light.

Primary xylem: Vascular bundles clearly separated. Vessels solitary and in short radial multiples. Fibres thin-walled in radial rows.

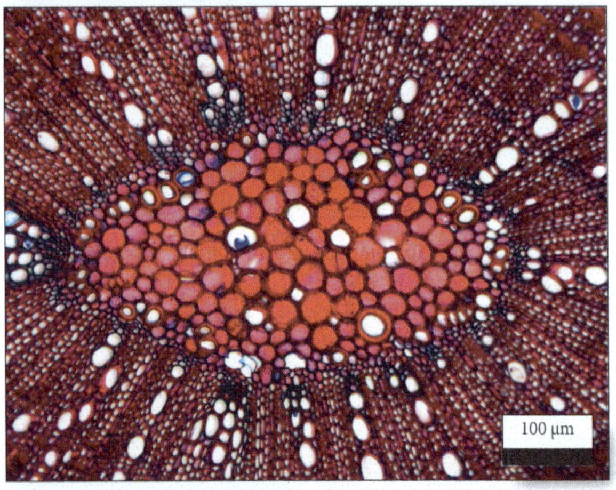

BRANCH XYLEM ANATOMY

Description: Growth ring boundaries distinct. Wood ring-porous. Tangential diameter of earlywood vessels 90-175-240 µm and latewood vessels 30-68-100 µm; 39-55-69 vessels/mm². Vessel element length 50-91-140 µm. Intervessel pits small, 4-7 µm in diameter. Fibres thin- and thick-walled. Axial parenchyma vasicentric and in bands 3(10) cells wide. Rays exclusively uniseriate. Ray height 40-105-160 µm, rays 5-7-10/mm.

Anatomical differences to stem xylem: Branch wood is ring-porous while stem wood is diffuse-porous, therefore quantitative comparison based on vessel features is not possible. Rays are similar in height comparing branch to stem (40-105-160/40-106-140 µm), while the ray number is lower in branch (7-10-14/11-13-17 rays/mm).

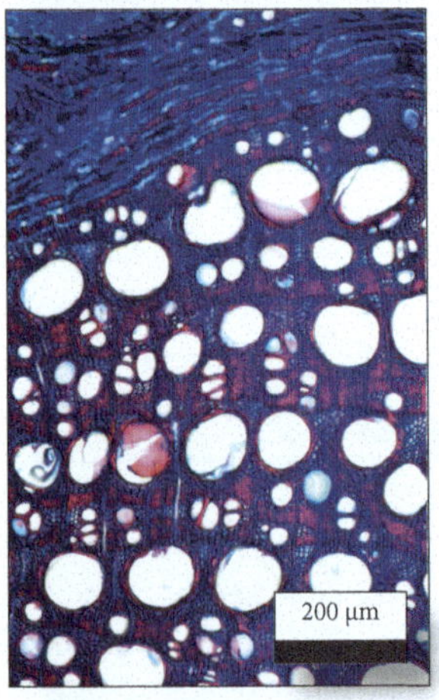

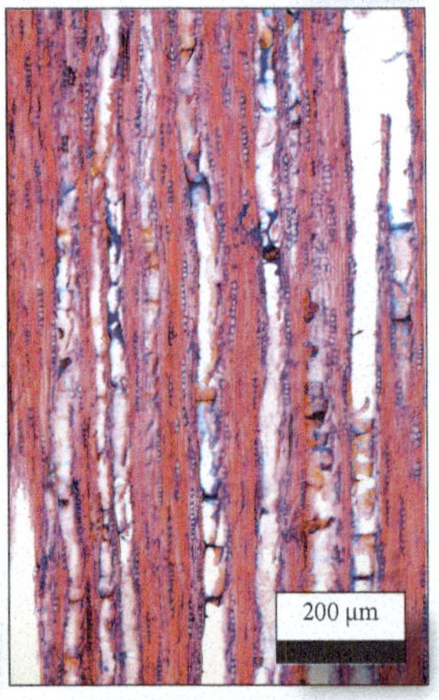

BARK ANATOMY

Branch bark: Collapsed sieve tubes alternating with groups of thick-walled fibres. The cortex consists of irregularly blue-stained parenchyma cells. Epidermis distinct in polarized light.

Stem bark: The phloem consists of alternating tangential bands of fibres and collapsed and non-collapsed sieve tubes. Axial parenchyma dilatation present. The layered phellem is homogeneous and it consists of regularly arranged rectangular cells. Lignified cells in phellem with prismatic crystals, distinct in polarized light.

IAWA feature numbers
1 5 9 10 13 22 25 29 30 41 48 52 61 66 69 77 79 80 82 90 91 96 102.2 106 116 118 120 136 142 179 180 191 192 196

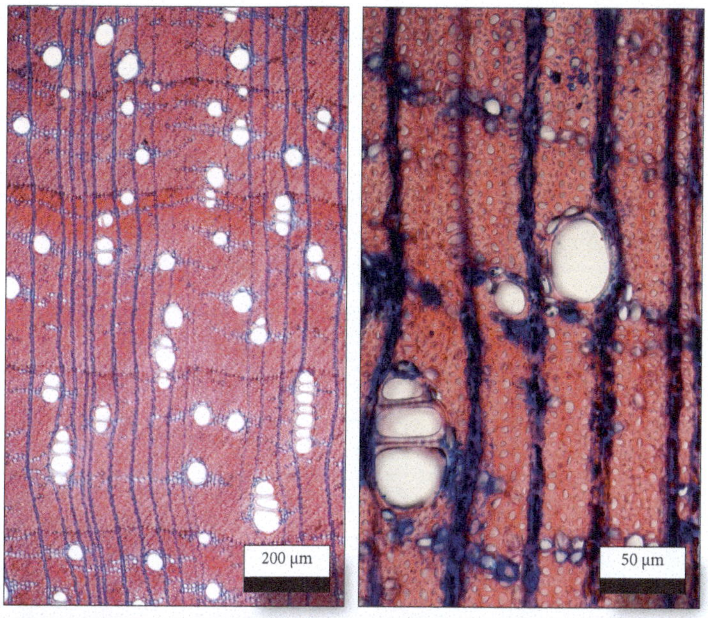

STEM XYLEM ANATOMY

Transverse section

Growth ring boundaries distinctly, marked by thick-walled radially flattened fibres.
Wood diffuse-porous.
Vessels solitary or in radial multiples of 2-7 elements. Tangential diameter of vessels 50-68-80 µm, 11-27-38 vessels/mm^2.
Fibres thin- to thick-walled.
Apotracheal parenchyma diffuse-in-aggregates. Paratracheal parenchyma vasicentric, in some vessels winged-aliform to aliform.
Rays 11-13-17/mm.

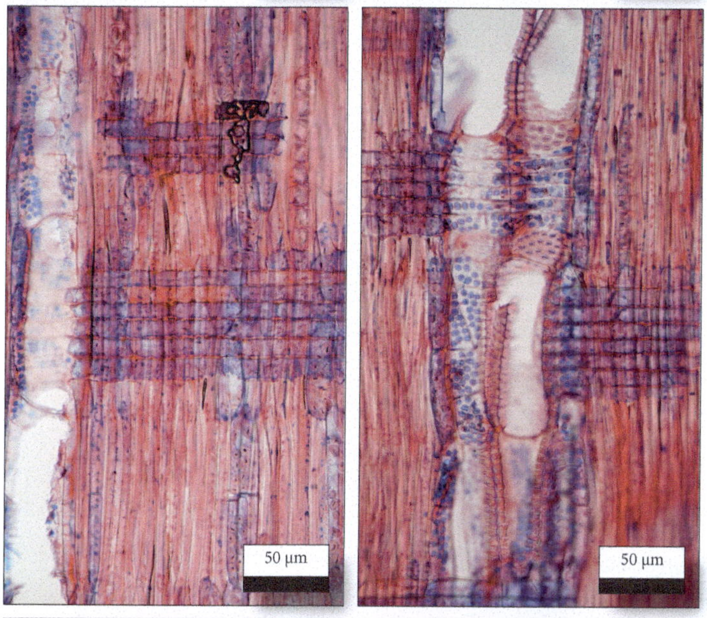

Radial section

Simple perforation plates.
Intervessel pits oval to circular and alternate, 4-7 µm in diameter.
Vestured pits present.
Vessel-ray pits with distinct borders; similar to intervessel pits in size and shape throughout the ray cell.
Vessel element length 110-158-190 µm.
Fibres with simple to minutely bordered pits (libriform fibres).
Heterogeneous rays with body ray cells procumbent and one row of square marginal cells.
Prismatic crystal in chambered axial parenchyma cells.

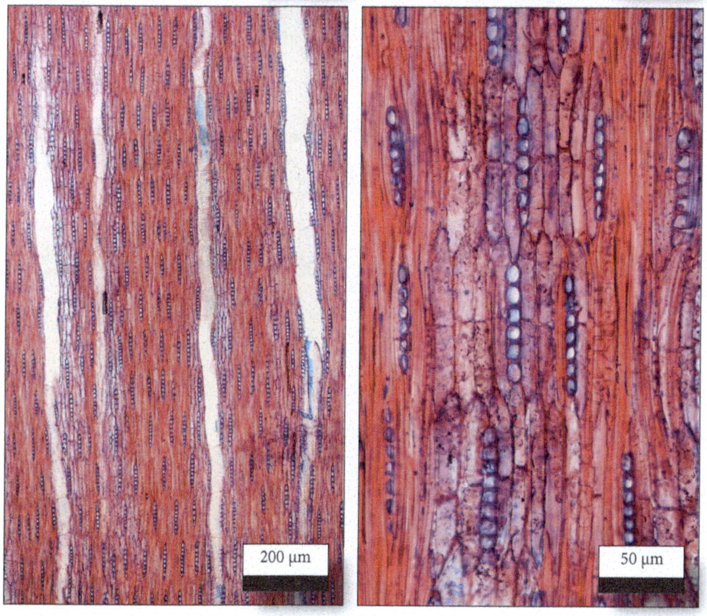

Tangential section

Rays exclusively uniseriate.
Ray height 40-106-140 µm.
Fusiform parenchyma cells present. Two cells per parenchyma strand.
Rays, vessels and axial parenchyma storied. Number of ray tiers per axial mm: 14-17.

Dalbergia chapelieri Baill.

Synonymy: *D. pterocarpiflora* Baker

EN: Palisander FR: Palissandre ML: Voamboana

PLANT DESCRIPTION

Shrub to tree | 4-15 m high | 11-22 cm DBH
Endemic to Madagascar. Exploited species.

DISTRIBUTION

Elevation: 0-1000 m a.s.l.
Vegetation formation: Degraded formation and forests
Bioclimate: Humid
Protected areas: Analamazaotra-Périnet, Andasibe, Andohahela, Andringitra, Betampona, Manombo, Midongy du Sud, Pic d'Ivohibe, Ranomafana and Kirindy-Menabe
New protected areas: Analalava, Sahafary and Tampolo

Slides made from material collected in Andasibe.

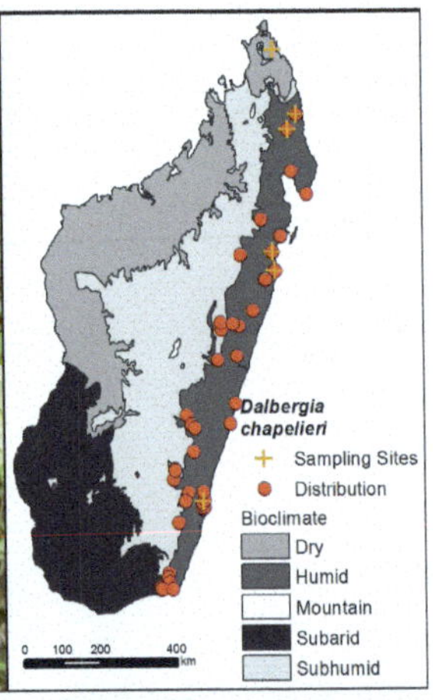

PITH AND PRIMARY XYLEM ANATOMY

Pith: Shape of pith round. A layer of blue-stained cells surrounds the pith. Cells dimorphic. Pith not visible in polarized light.

Primary xylem: Vascular bundles not distinct. Vessels solitary and in long radial multiples.

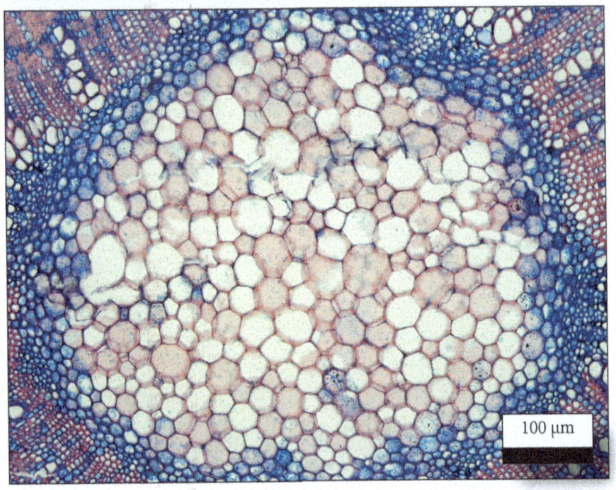

BRANCH XYLEM ANATOMY

Description: Growth rings boundaries indistinct or absent. Tangential diameter of vessel 80-115-180 µm, 2-8-20 vessels/mm². Vessel element length 130-178-290 µm. Intervessel pits small, 4-7 µm in diameter. Fibres thin- to thick-walled. Axial parenchyma diffuse in some vessels winged-aliform. Apotracheal parenchyma in narrow bands. Rays exclusively uniseriate, ray height 110-145-260 µm. Rays 9-11-14/mm.

Anatomical differences to stem xylem: Mean tangential diameter of vessel lumina is slightly smaller in the branch than in stem (80-115-180/80-122-170 µm) while number of vessel is similar (2-8-20/3-7-13 vessels/mm²). Rays are lower in the branch (110-145-260/50-161-310 µm) while ray number is similar (9-11-14/8-10-14 rays/mm). Axial elements not storied in the branch, while rays, vessels and axial parenchyma are irregularly storied in the stem.

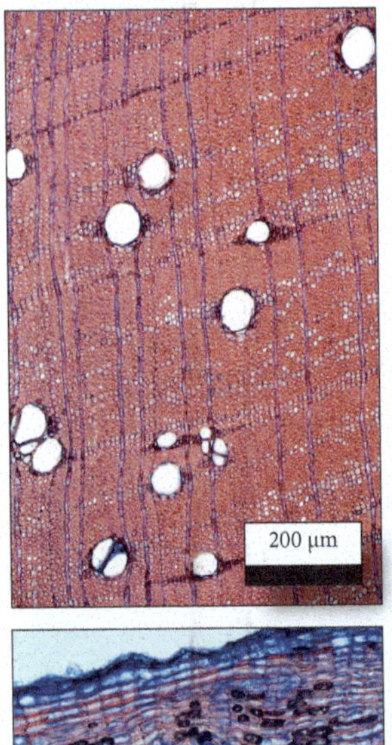

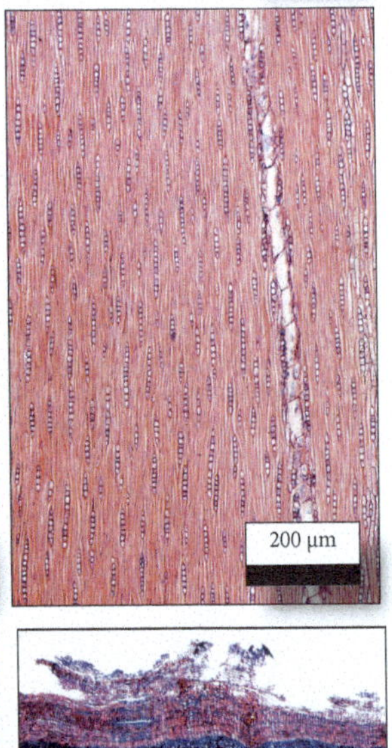

BARK ANATOMY

Branch bark: The phloem consists of alternating groups of fibres and collapsed and non-collapsed sieve tubes. The cortex consists of irregular round and oval blue-stained cells.

Stem bark: The phloem consists of alternating bands of fibres and collapsed and non-collapsed sieve tubes. Distinct ray and parenchyma dilatations present. Only some rays become dilated. Crystal sand irregularly dispersed in phloem. The phellem is homogeneous, it consists of regularly arranged rectangular cells. Lignified cells in phellem, distinct in polarized light.

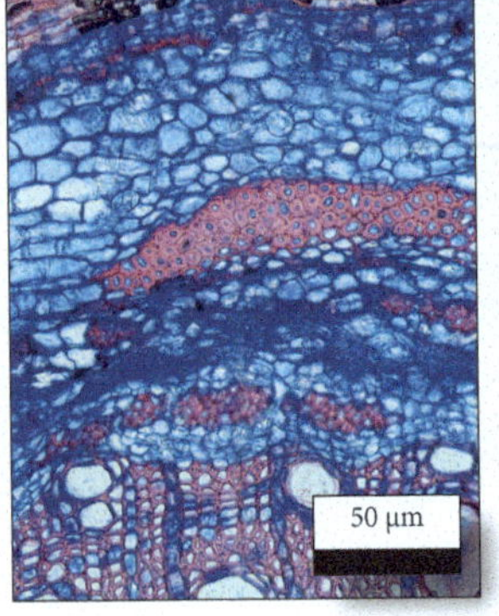

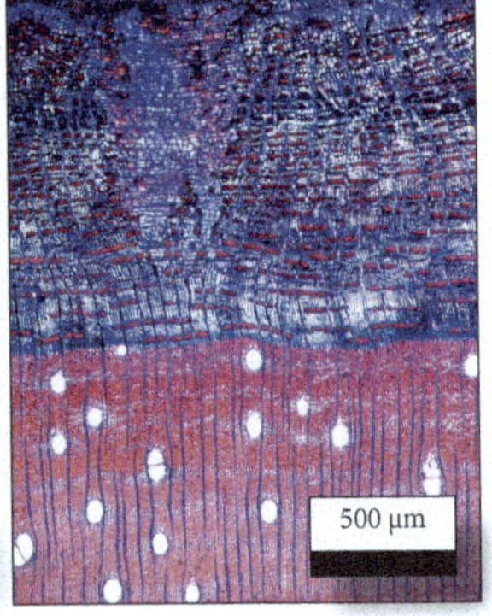

IAWA feature numbers

2 5 9 13 22 25 29 30 42 47 52 61 66 70 76 77 78 79 90 91 96 102.2 104 116 118 120 136 142 178 180 189 190 192 196

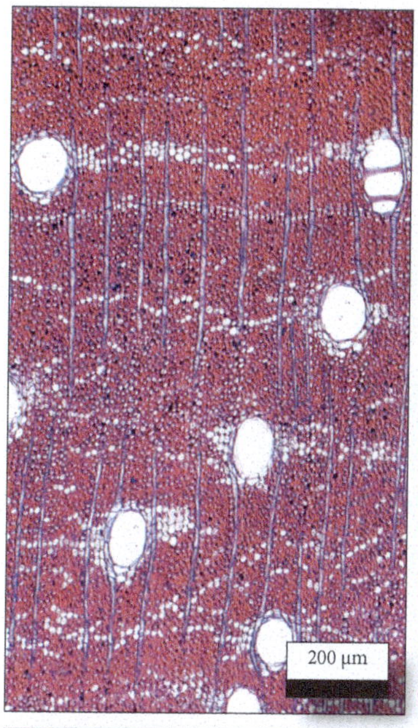

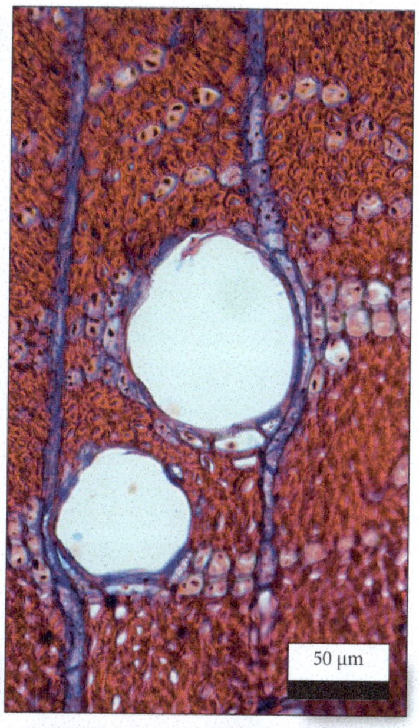

STEM XYLEM ANATOMY

Transverse section

Growth ring boundaries indistinct or absent.
Wood diffuse-porous.
Vessels predominantly solitary.
Tangential diameter of vessel 80-122-170 μm, 3-7-13 vessels/mm^2.
Fibres very thick-walled.
Apotracheal parenchyma diffuse and diffuse-in-aggregates. Paratracheal parenchyma scanty to vasicentric.
Rays 8-10-14/mm.

Radial section

Simple perforation plates.
Intervessel pits oval to circular and alternate, 4-7 μm in diameter.
Vestured pits present.
Vessel-ray pits with distinct borders; similar to intervessel pits in size and shape throughout the ray cell.
Vessel element length 160-217-270 μm.
Fibres with simple to minutely bordered pits.
Homogeneous rays with all cells procumbent.
Prismatic crystal in chambered axial parenchyma cells.

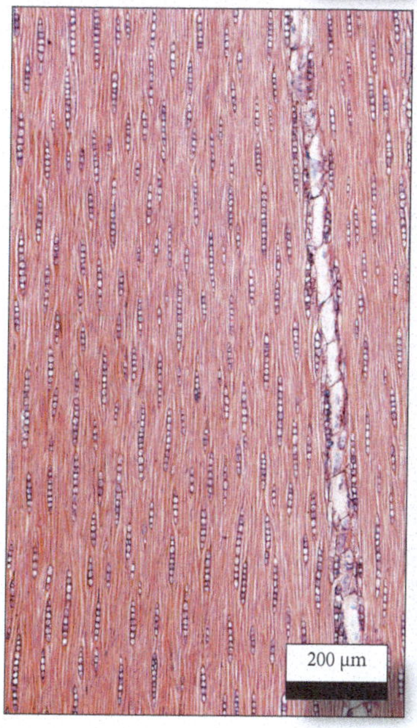

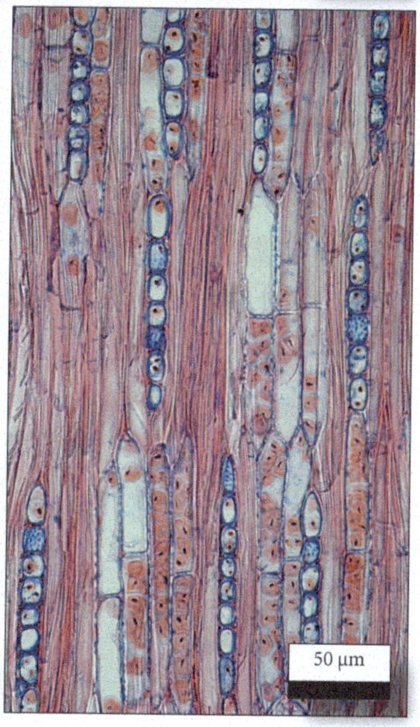

Tangential section

Rays exclusively uniseriate.
Ray height 50-161-310 μm.
Fusiform parenchyma cells present. Two cells per parenchyma strand.
Rays, vessels and axial parenchyma irregularly storied.
Number of ray tiers per axial mm: 10-11.

Dalbergia emirnensis Benth.

EN: Palisandre FR: Palissandre ML: Tsiandalana

PLANT DESCRIPTION

Shrub to tree | 11-15 m high
Endemic to Madagascar. Exploited species.

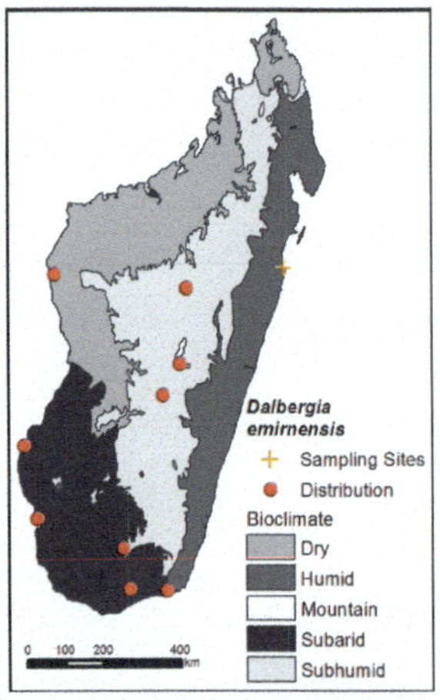

DISTRIBUTION

Elevation: 0-1500 m a.s.l.
Vegetation formation: Forests, freshwater wetland, inselberg/rock face and thicket
Bioclimate: Dry, sub-arid and sub-humid
Protected areas: Andohahela
New protected areas: Montagne des Français and Itremo/ Ambatofinandrahana

Slides made from material collected in Montagne des Français.

PITH AND PRIMARY XYLEM ANATOMY

Pith: Shape of pith round. A layer of blue-stained cells surrounds the pith. Thin-walled parenchyma cells present. Cells dimorphic. Prismatic crystals present. Simple pits in transverse cell walls. Pith visible in polarized light.

Primary xylem: Vascular bundles not distinct. Vessels solitary and in short radial multiples.

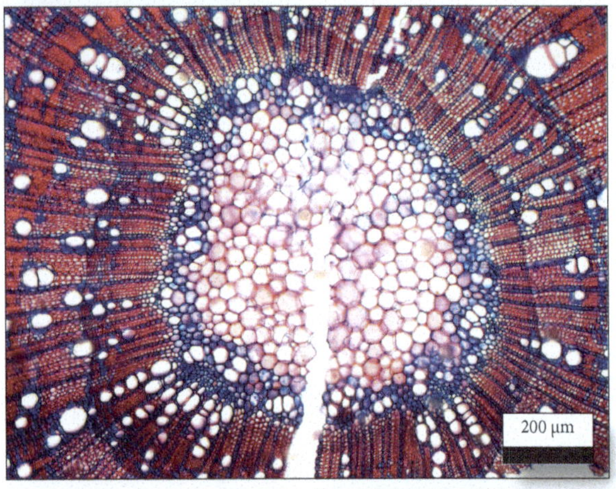

BRANCH XYLEM ANATOMY

Description: Growth ring boundaries absent or indistinct. Tangential diameter of vessels 60-111-160 µm, 12-17-23 vessels/mm². Vessel element length 180-209-230 µm. Intervessel pits medium, 7-10 µm in diameter. Fibres thin- to thick-walled. Axial parenchyma in bands 2-9 cells wide. Rays uni- and biseriate, predominantly uniseriate. Ray height 40-125-260 µm. Rays 8-10-12/mm.

Anatomical differences to stem xylem: Mean tangential diameter of vessel lumina is smaller in the branch than in the stem (60-111-160/90-179-250 µm). Vessel density is lower in the branch (12-17-23/17-28-35 vessels/mm²). Rays are slightly higher in the stem (40-125-260/120-173-320 µm) while ray number is similar (8-10-12/9-11-14 rays/mm).

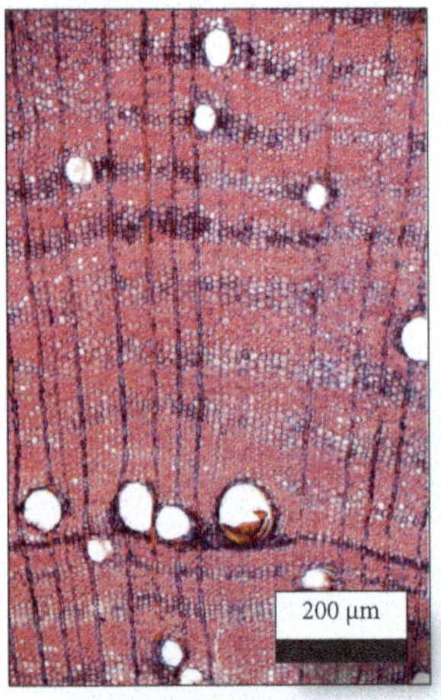

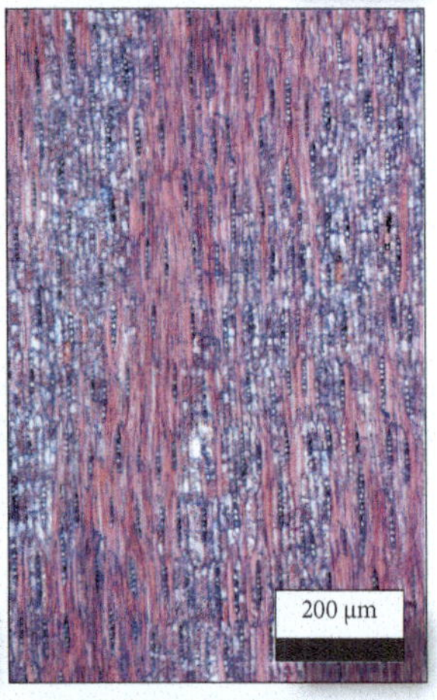

BARK ANATOMY

Branch bark: Collapsed sieve tubes present in the phloem. The cortex consists of round and oval blue-stained parenchyma cells with groups of tangentially distributed thick-walled fibres.

Stem bark: The phloem consists of alternating tangential bands of fibres and collapsed and non-collapsed sieve tubes. Axial parenchyma dilatation present. The layered phellem is homogeneous and it consists of regularly arranged rectangular parenchyma cells with dark staining substances. Lignified cells in phellem, distinct in polarized light.

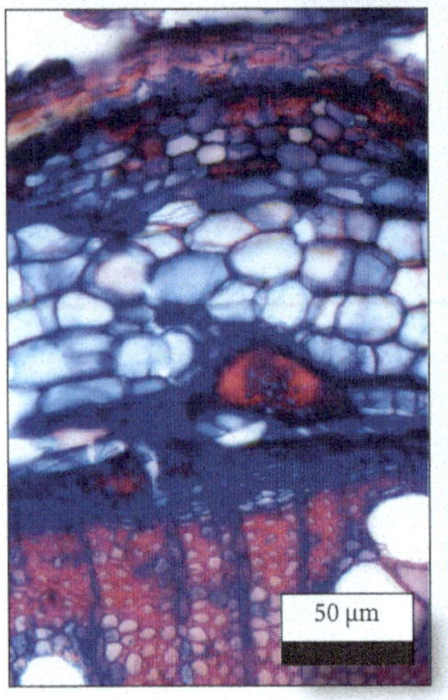

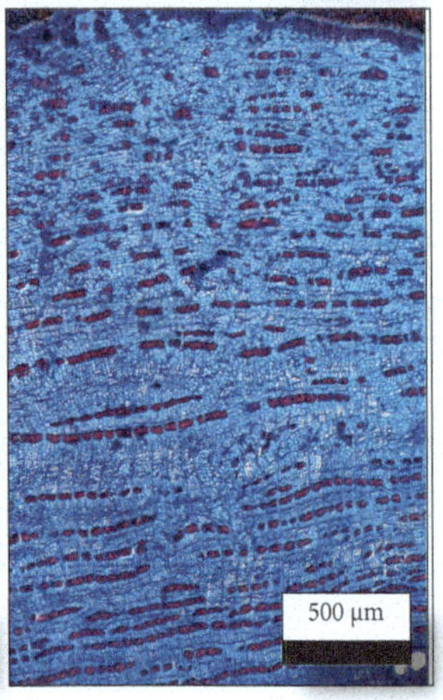

IAWA feature numbers
2 5 9 9.1 13 22 26 29 30 42 48 52 58 61 66 69 79 85 90 91 97 102.2 104 116 118 120 121 136 142 178 180 189 190 192 196

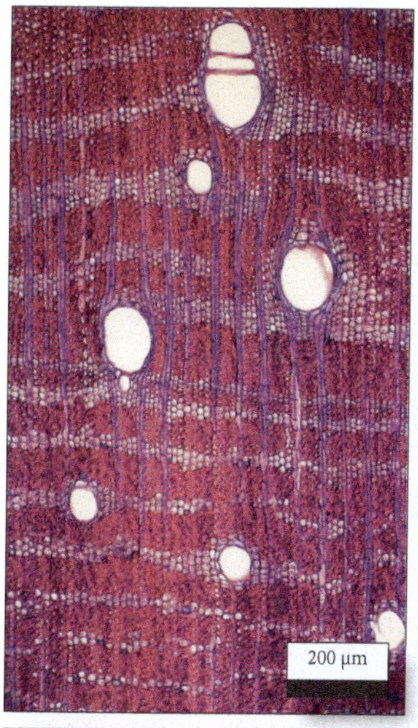

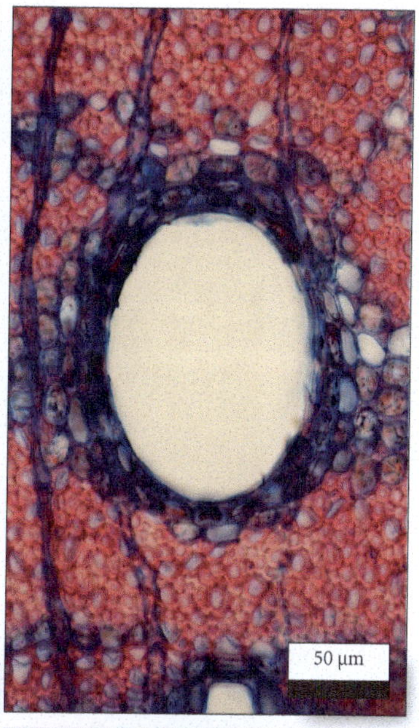

STEM XYLEM ANATOMY

Transverse section
Growth ring boundaries indistinct or absent.
Wood diffuse-porous.
Vessels mostly solitary. A few short radial multiples of 2-4 cells present.
Tangential diameter of vessels large 90-179-250 µm, 17-28-35 vessels/mm^2.
Gums and/or other deposits in heartwood vessels.
Fibres thin to thick-walled.
Paratracheal parenchyma vasicentric.
Apotracheal parenchyma in bands more than three cells wide.
Rays 9-11-14/mm.

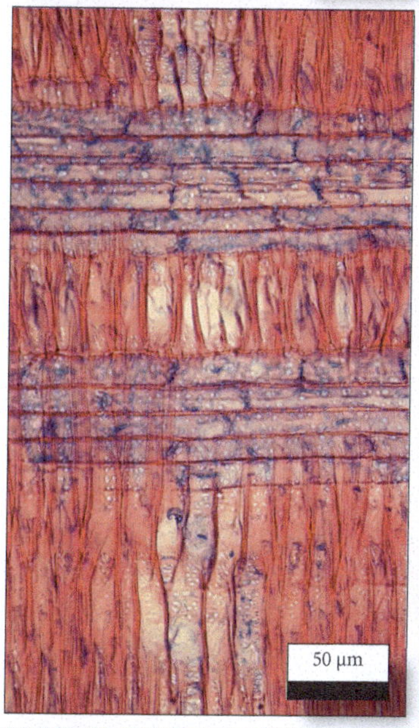

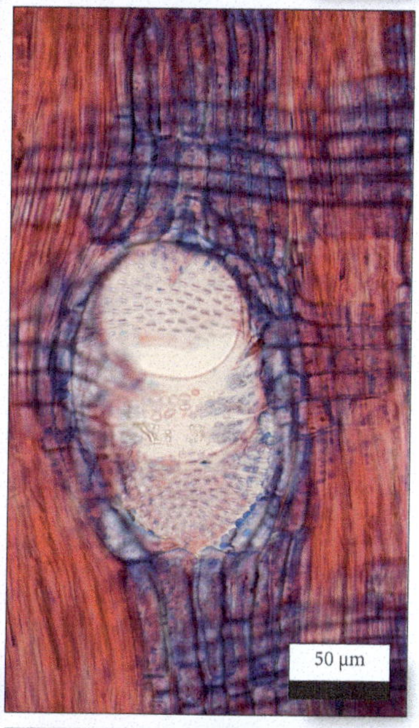

Radial section
Simple perforation plates.
Intervessel pits oval to circular and alternate, 6-8 µm in diameter.
Vestured pits present.
Vessel-ray pits with distinct borders; similar to intervessel pits in size and shape throughout the ray cell.
Vessel element length 180-191-210 µm.
Fibres with simple to minutely bordered pits.
Homogeneous rays with all cells procumbent.
Prismatic crystal in chambered axial parenchyma cells.

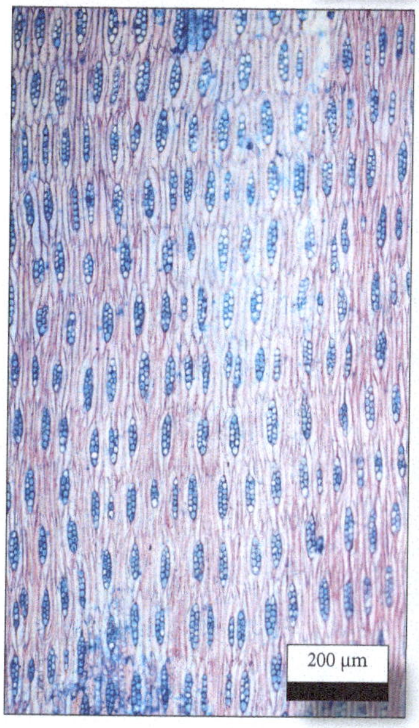

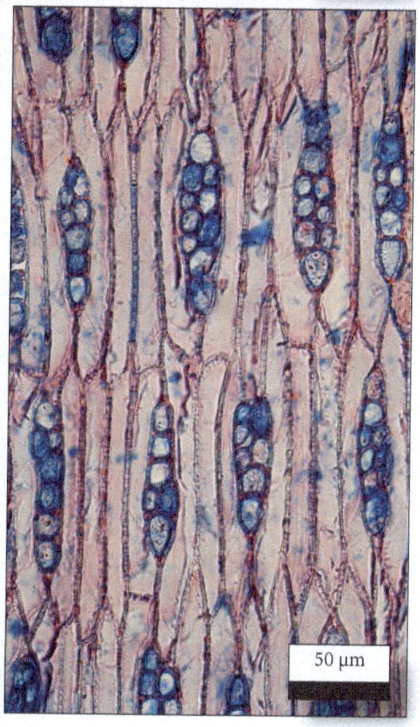

Tangential section
Rays predominantly biseriate.
Ray height 120-173-320 µm.
Axial parenchyma strands composed of 1-2 cells.
Fusiform parenchyma cells present. Two cells per parenchyma strand.
Rays, vessels and axial parenchyma cells storied.
Number of ray tiers per axial mm: 13-14.

Dalbergia greveana Baill.

Synonymy: *D. ambongoensis* Baill., *D. eurybothrya* Drake., *D. myriabotrys* Baker., *D. perrieri* Jum., *D. ikopensis* Jum., *D. isaloensis* R.Vig.

EN: Palisander FR: Palissandre ML: Manary

PLANT DESCRIPTION

Shrub to tree | 3-20 m high | 15-80 cm DBH

Endemic to Madagascar. Exploited species.

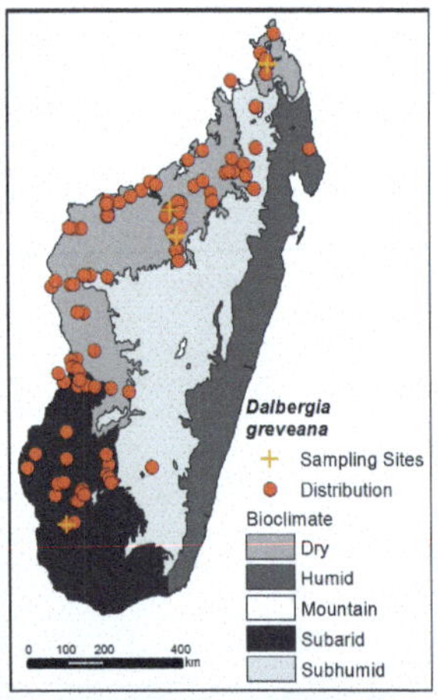

DISTRIBUTION

Elevation: 0-1000 m a.s.l.

Vegetation formation: Forest, grassland and woodland

Bioclimate: Dry, sub-arid and sub-humid

Protected areas: Ankarafantsika, Ankarana, Isalo, Bemaraha, Kirindy-Mitea, Namoroka, Tsaratanana and Zombitsy-Vohibasia

New protected areas: Amoron'i Onilahy and Montagne des Français

Slides made from material collected in Ankarana.

PITH AND PRIMARY XYLEM ANATOMY

Pith: Shape of pith round. Thick-walled parenchyma cells present. Cells dimorphic. Prismatic crystals present. Simple pits in transverse cell walls. Pith not visible in polarized light.

Primary xylem: Vascular bundles separated. Vessels solitary and in short radial multiples.

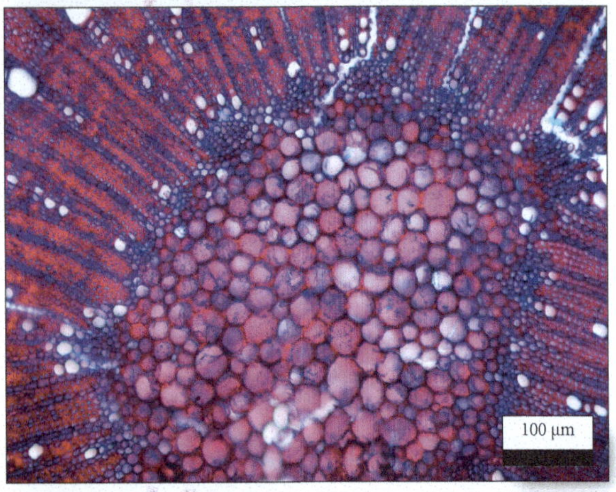

BRANCH XYLEM ANATOMY

Description: Growth ring boundaries indistinct. Wood diffuse-porous. Tangential diameter of earlywood vessels medium 70-92-130 µm, 20-38-61 vessels/mm². Vessel elements length 130-142-160 µm. Intervessel pits small, 4-7 µm in diameter. Fibre thick- to very thick-walled. Axial parenchyma diffuse and diffuse-in-aggregates. Rays predominantly uniseriate. Ray height 60-86-130 µm, rays 6-7-9 rays/mm.

Anatomical differences to stem xylem: Branch wood is ring-porous while stem wood is diffuse-porous, therefore quantitative comparison based on vessel features is not possible. Rays are shorter in the branch (60-86-130/50-145-260 µm), while ray number is similar (6-7-9/4-6-9 rays/mm).

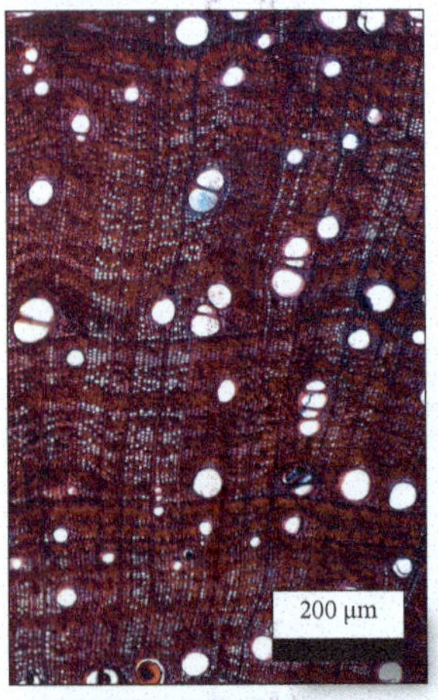

BARK ANATOMY

Branch bark: Collapsed and non-collapsed sieve tubes present in the phloem. The cortex consists of irregularly blue-stained parenchyma cells and groups of fibres in tangential bands.

Stem bark: The phloem consists of alternating tangential bands of fibres and collapsed and non-collapsed sieve tubes. Axial parenchyma dilatation present. The phellem is homogeneous, it consists of regularly arranged rectangular cells. Lignified cells in phellem, distinct in polarized light.

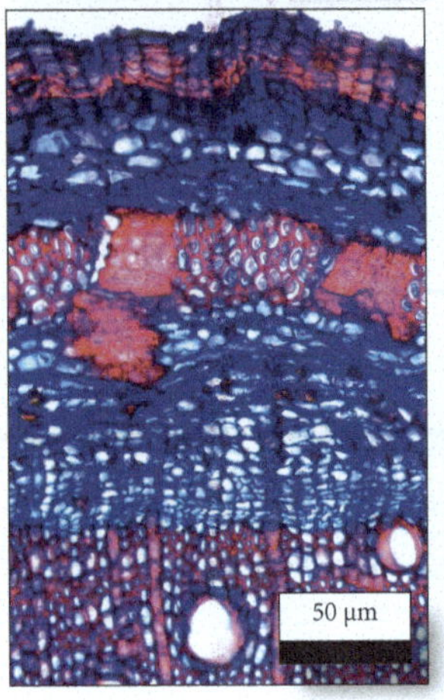

IAWA feature numbers

1 4 9 9.1 13 22 25 29 30 42 47 52 58 61 66 69 77 79 90 91 97 104 115 118 119 120 122 136 138 142 180 189 196

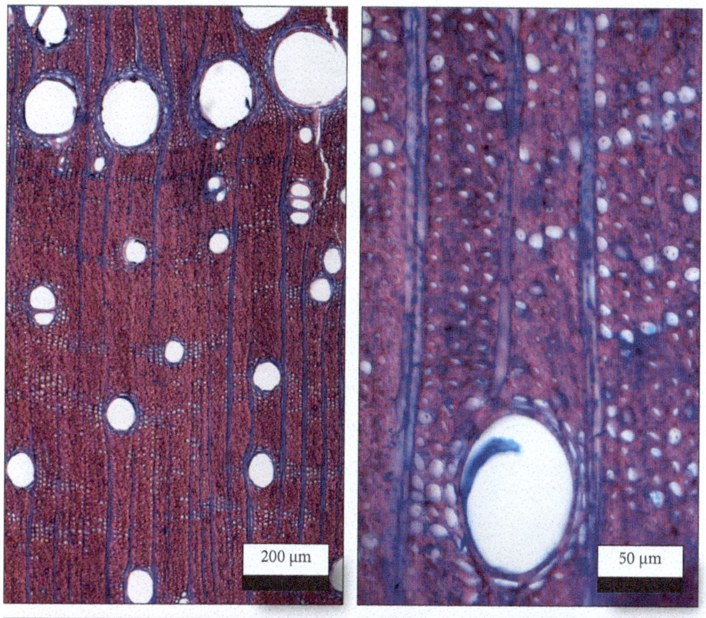

STEM XYLEM ANATOMY

Transverse section

Growth ring boundaries distinct, marked by thick-walled and radially flattened latewood fibres and by differences in vessels diameter.
Wood semi-ring-porous.
Vessels solitary and in short radial multiples. Tangential diameter of earlywood vessels 140-177-240 µm and latewood vessel 50-89-120 µm, 13-18-28 vessels/mm².
Gums and/or other deposits in heartwood vessels.
Fibres thin- to thick-walled.
Apotracheal parenchyma diffuse-in-aggregates.
Paratracheal parenchyma vasicentric.
Rays 4-6-9/mm.

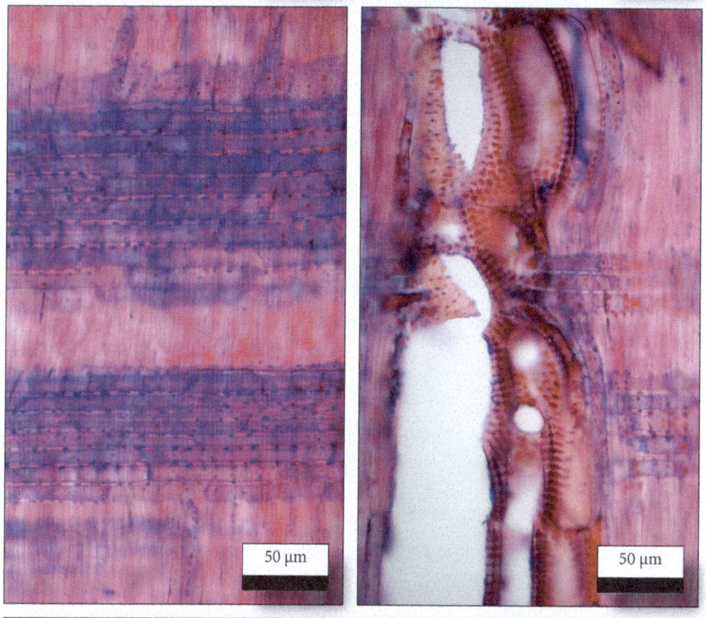

Radial section

Simple perforation plates.
Intervessel pits oval to circular and alternate, 4-7 µm in diameter.
Vestured pits present.
Vessel-ray pits with distinct borders; similar to intervessel pits in size and shape throughout the ray cell.
Vessel element length 120-154 -180 µm.
Fibres with simple to minutely bordered pits.
Homogeneous rays with all cells procumbent.
Prismatic crystal in procumbent ray cells and in chambered axial parenchyma cells.

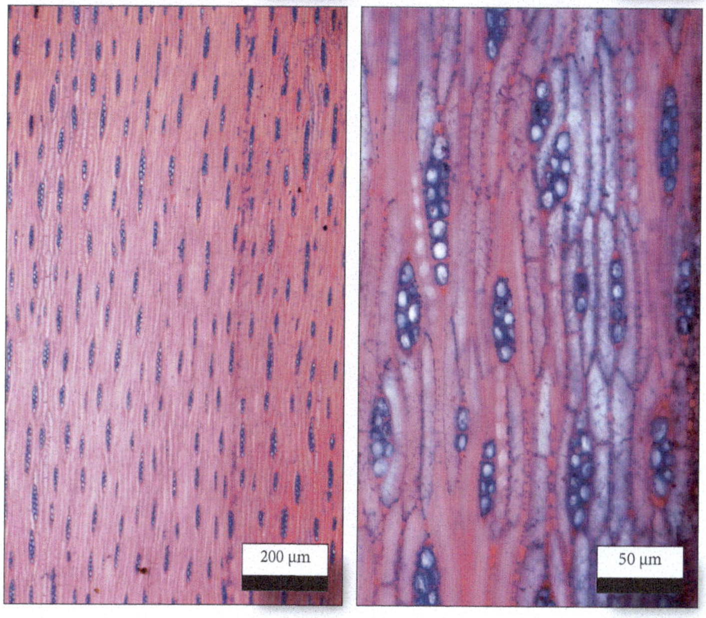

Tangential section

Rays predominantly biseriate. Ray height 50-145-260 µm.
Fusiform parenchyma cells present. Two cells per parenchyma strand.
Rays and axial parenchyma irregularly storied.
Number of ray tiers per axial mm: 15-16.

Dalbergia lemurica Bosser and R. Rabev.

EN: Palisander FR: Palissandre ML: Manary

PLANT DESCRIPTION

Tree | 6-10 m high | 20-100 cm DBH
Endemic to Madagascar. Exploited species.

DISTRIBUTION

Elevation: 0-500 m a.s.l.
Vegetation formation: Forest and woodland
Bioclimate: Dry and sub-arid
Protected areas: Bemaraha, Kirindy-Mitea
New protected areas: Orangea/Rigny and Sahafary

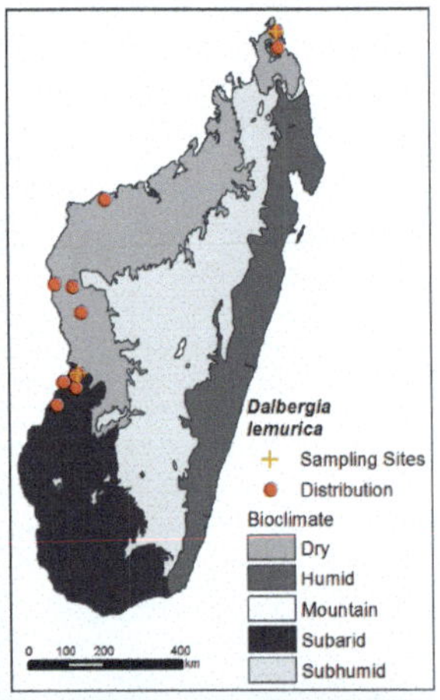

Slides made from material collected in Orangea/Rigny.

PITH AND PRIMARY XYLEM ANATOMY

Pith: Shape of pith polygonal. Prismatic crystals rare. Simple pits in transverse cell walls. Pith visible in polarized light.

Primary xylem: Vascular bundles not distinct. Vessels solitary and in short radial multiples.

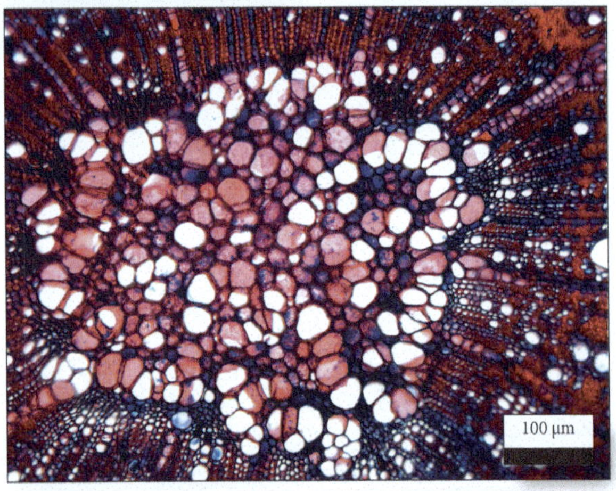

BRANCH XYLEM ANATOMY

Description: Growth rings boundaries indistinct or absent. Tangential diameter of vessels 50-87-180 µm, 7-15-24 vessels/mm^2. Vessel element length 130-157-180 µm. Intervessel pits small, 4-7 µm in diameter. Fibres very thick-walled. Axial parenchyma vasicentric, diffuse-in-aggregates and in bands up to 3 cells. Rays predominantly uniseriate. Ray height 40-114-170 µm. Rays 9-12-14/mm.

Anatomical differences to stem xylem: Mean tangential diameter of vessel lumina is slightly higher in the stem (50-87-180/60-94-140 µm). Vessel density is higher in the branch (7-15-24/3-6-10 vessels/mm^2). Rays are slightly higher in the branch (40-114-170/70-107-130 µm) while ray number is similar (9-12-14/11-12-15 rays/mm).

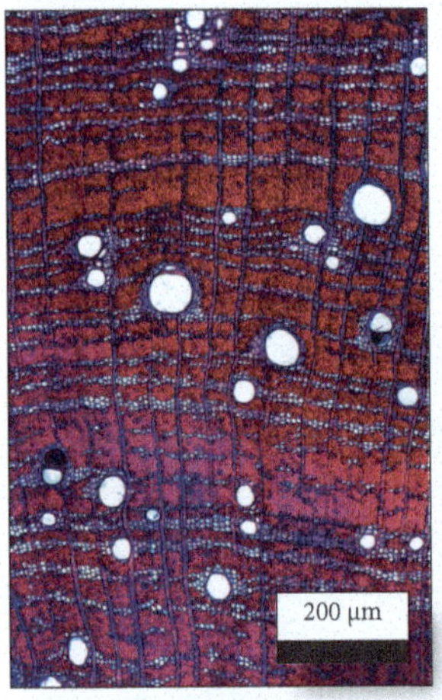

BARK ANATOMY

Branch bark: Sieve tubes, groups of fibre and collapsed sieve tubes present in the phloem. The cortex consists of large blue-stained parenchyma cells and group of tangentially distributed fibres.

Stem bark: The phloem consists of alternating tangential bands of fibres and collapsed and non-collapsed sieve tubes. Axial parenchyma dilatation present. The layered phellem is homogeneous and consists of regularly arranged rectangular parenchyma cells. Lignified cells in phellem distinct in polarized light.

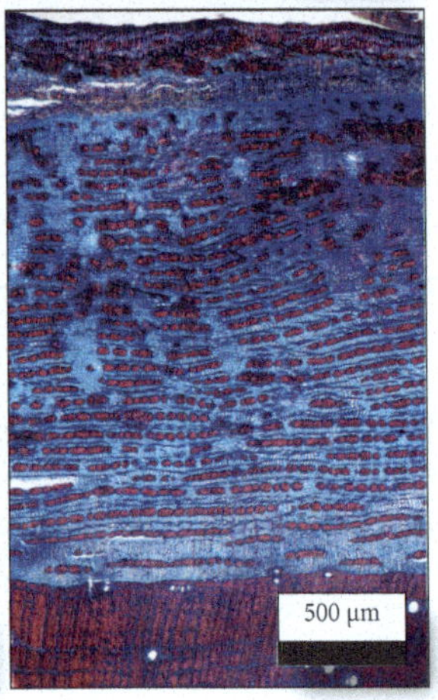

IAWA feature numbers

2 5 9 13 22 26 29 30 41 45 47 52 58 61 66 70 77 79 80 86 90 91 97 102.2 104 106 115 118 120 136 142 178 180 189 196

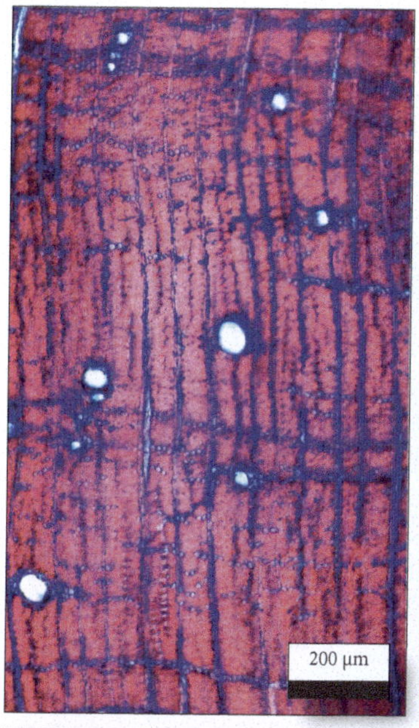

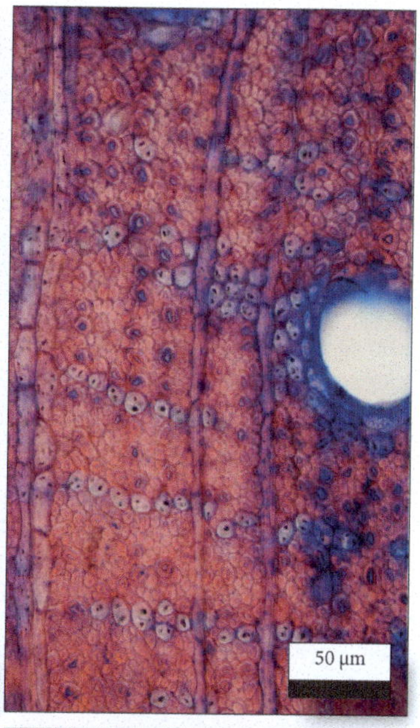

STEM XYLEM ANATOMY

Transverse section

Growth ring boundaries indistinct or absent.
Wood diffuse-porous.
Vessels mostly solitary.
Tangential diameter of vessels 60-94-140 µm, 3-6-10 vessels/mm^2.
Vessels of two distinct diameter classes.
Gums and/or other deposits in heartwood vessels.
Fibres very thick-walled.
Apotracheal parenchyma diffuse-in-aggregates and in narrows lines up to 3 cells. Paratracheal parenchyma vasicentric to aliform.
Rays 11-12-15/mm.

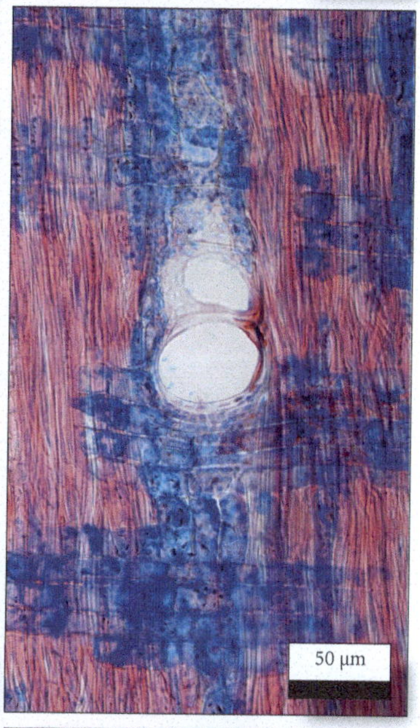

Radial section

Simple perforation plates.
Intervessel pits oval to circular and alternate, 7-10 µm in diameter.
Vestured pits present.
Vessel-ray pits with distinct borders; similar to intervessel pits in size and shape throughout the ray cell.
Vessel element length 80-158-210 µm.
Fibres with simple to minutely bordered pits.
Homogeneous and heterogeneous rays both present. In some rays body ray cells procumbent and one row of upright and square marginal cells.
One prismatic crystal per chamber in axial parenchyma cells.

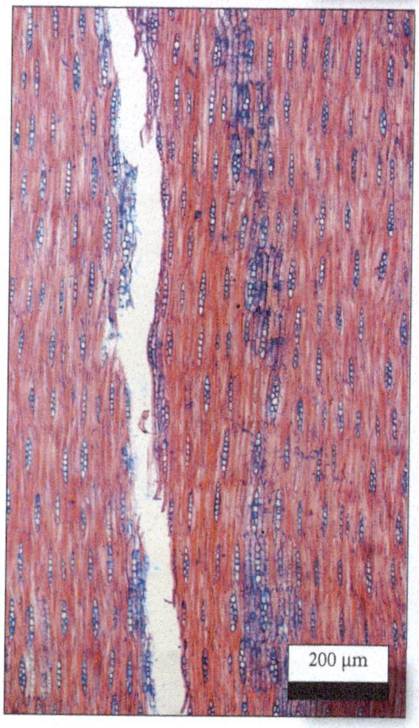

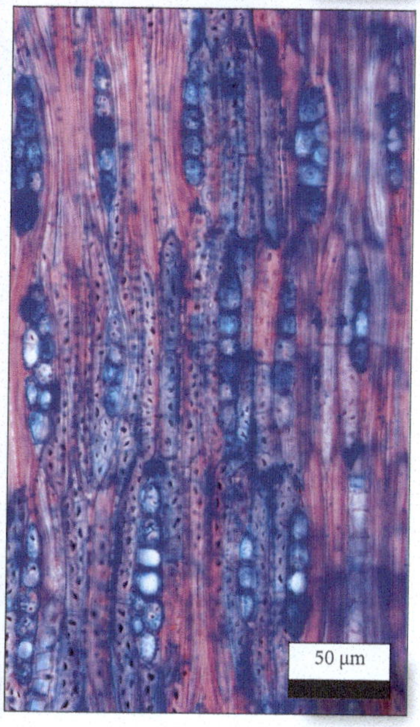

Tangential section

Rays predominantly biseriate. Ray height 70-107-130 µm.
Fusiform parenchyma cells present. Two cells per parenchyma strand.
Rays, vessels and axial parenchyma storied. Number of ray tiers per axial mm: 15-16.

Dalbergia louvelii R.Vig.

EN: Rosewood FR: Bois de rose ML: Andramena, Bolabola

PLANT DESCRIPTION

Tree | 8-10 m high | up to 18 cm DBH
Endemic to Madagascar. Exploited species.

DISTRIBUTION

Elevation: 0-1000 m a.s.l.
Vegetation formation: forest
Bioclimate: humid
Protected areas: Betampona and Mananara-Nord
New protected areas: Mahabo-Mananivo

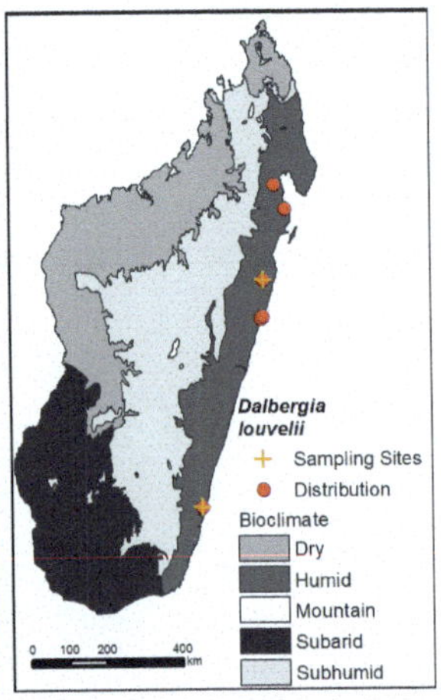

Slides made from material collected in Betampona.

PITH AND PRIMARY XYLEM ANATOMY

Pith: Shape of pith round to star-like. Cells dimorphic. Simple pits in transverse cell walls. Pith not visible in polarized light.

Primary xylem: Vascular bundles clearly separated. Vessels solitary and in short radial multiples. Fibres thin-walled in radial rows.

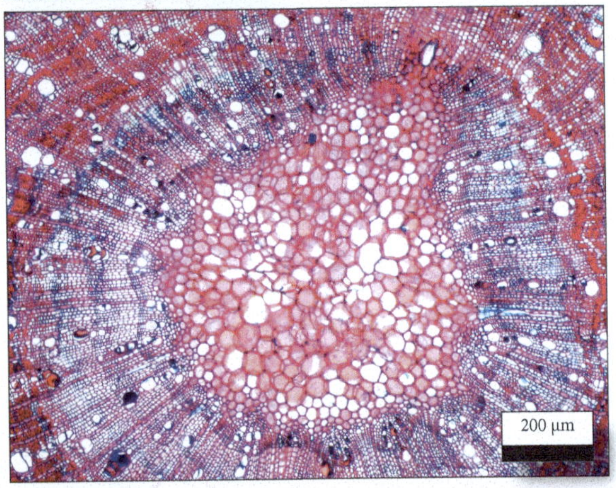

BRANCH XYLEM ANATOMY

Description: Growth ring boundaries indistinct. Tangential diameter of vessels 50-90-120 µm, 9-17-22 vessels/mm^2. Vessel element length 140-184-230 µm. Intervessel pits small, 4-7 µm in diameter. Fibres thick- to very thick-walled. Axial parenchyma paratracheal, diffuse-in-aggregates, and in narrow lines up to 3 cells. Rays exclusively uniseriate. Ray height 40-114-180 µm, rays 8-11-16/mm.

Anatomical differences to stem xylem: Mean tangential diameter of vessel lumina is smaller in the branch (50-90-120/110-161-260 µm). Vessel density is higher in the branch (9-17-22/5-13-20 vessels/mm^2). Rays are slightly shorter in the branch (40-114-180/50-190-410 µm) while ray number is similar (8-11-16/9-12-15 rays/mm).

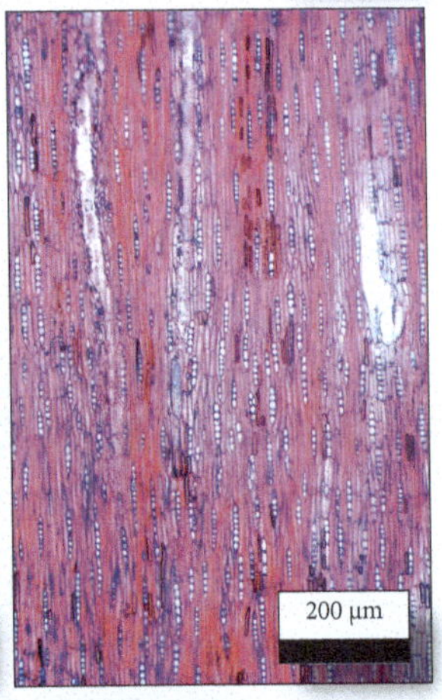

BARK ANATOMY

Branch bark: Non-collapsed and collapsed sieve tubes with two or three fibre cells per group present. The cortex consists of irregularly shaped blue-stained parenchyma cells and groups of tangentially scattered or irregularly dispersed fibres.

Stem bark: The phloem consist of alternating bands of fibre and collapsed and non-collapsed sieve tubes. Phellem homogeneous consists of regularly arranged rectangular cells. Lignified cells in phellem, distinct in polarized light.

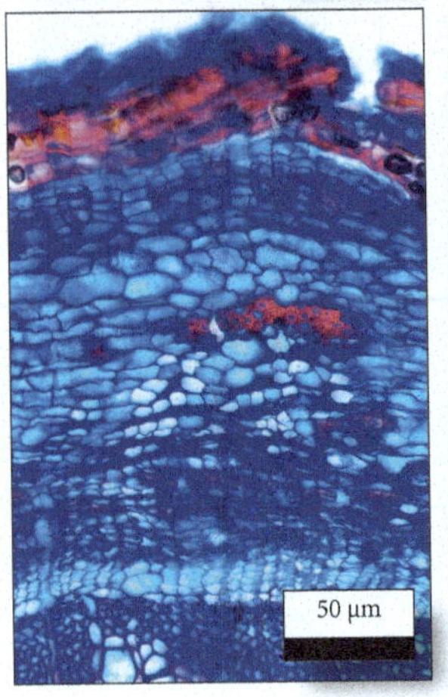

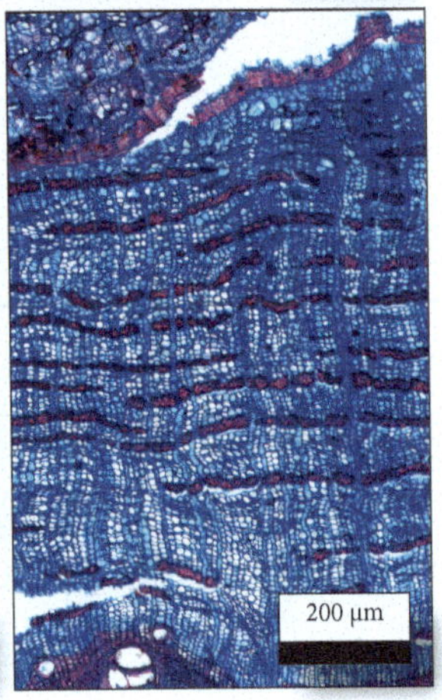

IAWA feature numbers
2 5 9 13 22 26 29 30 42 45 47 52 58 61 66 70 77 79 86 90 91 96.1 102.2 104 106 115 118 120 136 142 178 180 189 196

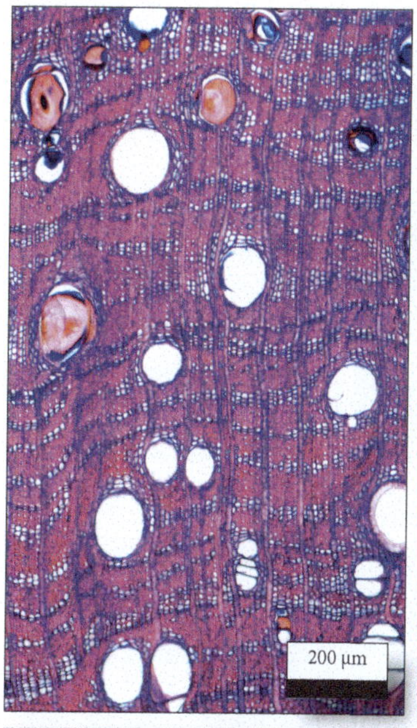

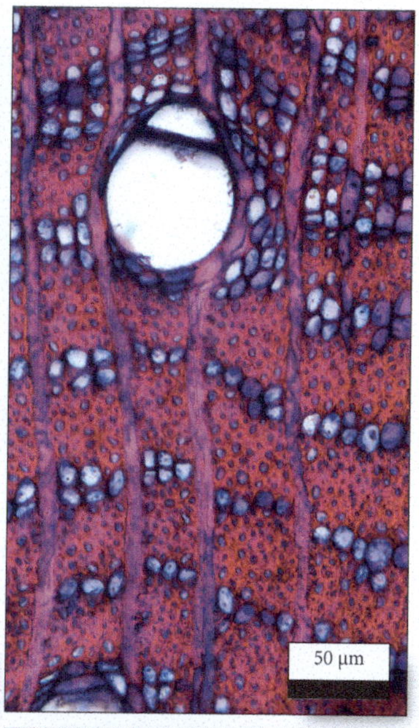

STEM XYLEM ANATOMY

Transverse section

Growth ring boundaries indistinct or absent.
Wood diffuse-porous.
Vessels predominantly solitary.
Tangential diameter of vessels 110-161-260 µm, 5-13-20 vessels/mm².
Vessels of two distinct diameter classes.
Gums and/or other deposits in heartwood vessels.
Fibres very thick-walled.
Paratracheal parenchyma scanty to vasicentric.
Apotracheal parenchyma diffuse-in-aggregates and in narrow bands up to three cells wide.
Rays 9-12-15/mm.

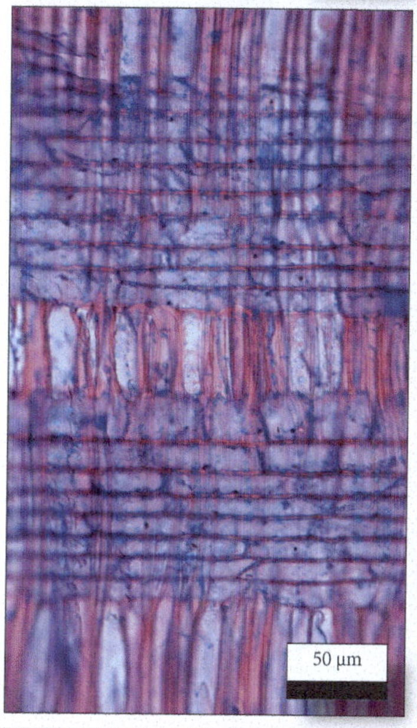

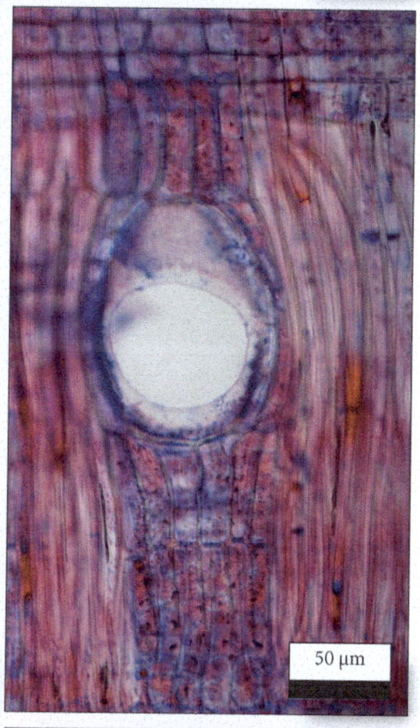

Radial section

Simple perforation plates.
Intervessel pits oval to circular and alternate, 7-10 µm in diameter.
Vestured pits present.
Vessel-ray pits with distinct borders; similar to intervessel pits in size and shape throughout the ray cell.
Vessel element length 90-162-330 µm.
Fibres with simple to minutely bordered pits.
Homogeneous and heterogeneous rays both present. In some rays body ray cells procumbent and one row of upright and square marginal cells.
Prismatic crystal in chambered axial parenchyma cells.

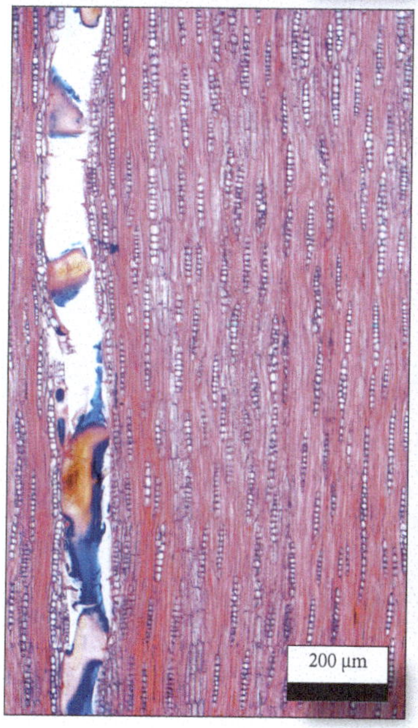

Tangential section

Rays predominantly uniseriate, some rays partially biseriate.
Ray height 50-190-410 µm.
Fusiform parenchyma cells present. Two cells per parenchyma strand.
Rays, vessels and axial parenchyma irregularly storied. Number of ray tiers per axial mm: 14-15.

Dalbergia madagascariensis Vatke.

EN: Palisander FR: Palissandre ML: Manary

PLANT DESCRIPTION

Tree | 5-20 m high | 20-45 cm DBH
Endemic to Madagascar. Exploited species.

DISTRIBUTION

Elevation: 0-1000 m a.s.l.
Vegetation formation: Forest and freshwater wetland
Bioclimate: Dry, humid and sub-humid
Protected areas: Ankarana, Betampona, Lokobe, Manongarivo, Marojejy and Montagne d'Ambre
New protected areas: Daraina complex, Montagne des Français, Makirovana, Analalava and Mahabo-Mananivo

Slides made from material collected in Makirovana.

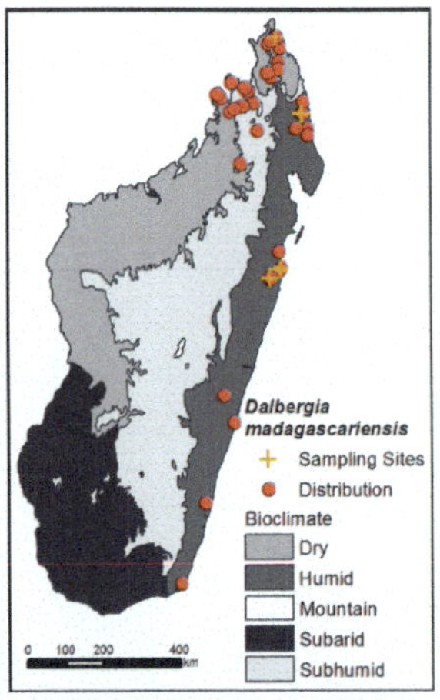

PITH AND PRIMARY XYLEM ANATOMY

Pith: Shape of pith square. Blue-stained cells present. Cells dimorphic. Prismatic crystals present. Simple pits in transverse cell walls. Pith not visible in polarized light.

Primary xylem: Vascular bundles clearly separated. Vessels solitary and in short radial multiples. Fibres thin- to thick-walled in radial rows.

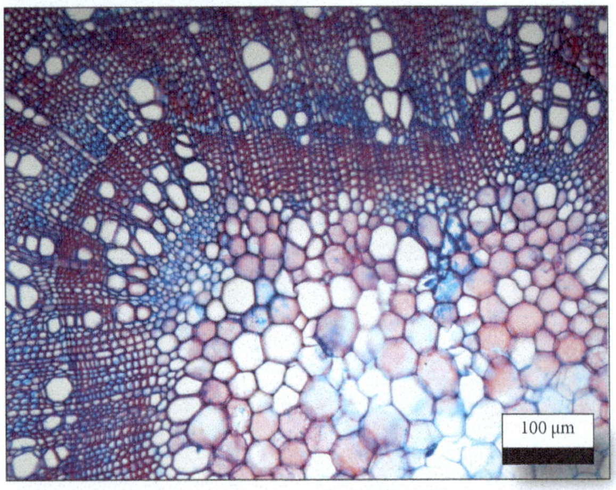

BRANCH XYLEM ANATOMY

Description: Growth rings boundaries indistinct or absent. Tangential diameter of vessels 100-175-220 µm, 9-15-20 vessels/mm^2. Vessel element length 90-100-130 µm. Intervessel pits small, 4-7 µm in diameter. Fibres thick- to very thick-walled. Axial parenchyma diffuse-in-aggregates and in narrow lines up to 3 cells. Rays exclusively uniseriate. Ray height 90-168-400 µm. Rays 6-8-12/mm.

Anatomical differences to stem xylem: Mean tangential diameter of vessel lumina is similar in the branch and in the stem (100-174-220/100-174-230 µm). Vessel number is similar in branch and stem (9-15-20/6-14-20 vessels/mm^2). Rays in the branch are slightly shorter (90-168-400/180-240-470 µm), while ray number is similar (6-8-12/ 7-8-11 rays/mm).

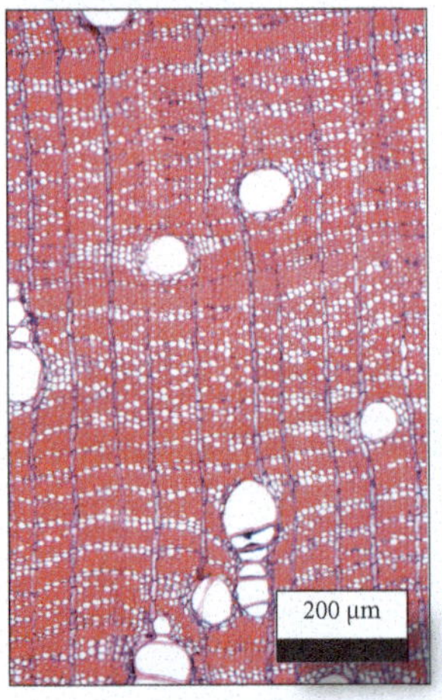

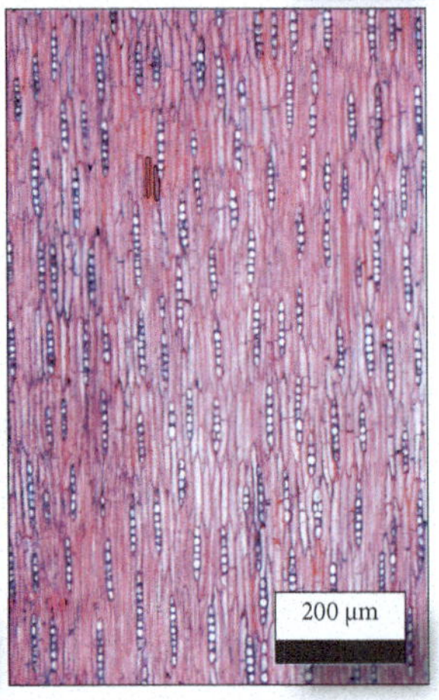

BARK ANATOMY

Branch bark: The phloem consists on alternating groups of fibres and collapsed and non-collapsed sieve tubes. The cortex consists of round and oval blue-stained parenchyma cells with some lignified cells irregularly distributed.

Stem bark: The phloem consists of alternating bands of fibres and collapsed and non-collapsed sieve tubes. Distinct ray and parenchyma dilatations present. Only some rays become dilated. Crystal sand irregularly dispersed in phloem. The phellem is homogeneous, it consists of regularly arranged rectangular cells. Lignified cells in phellem, distinct in polarized light.

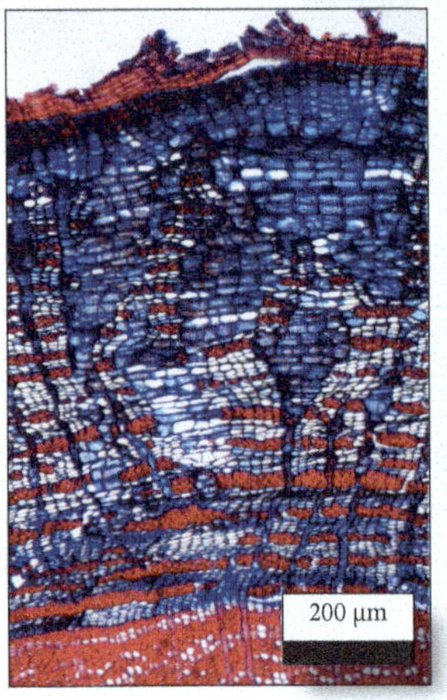

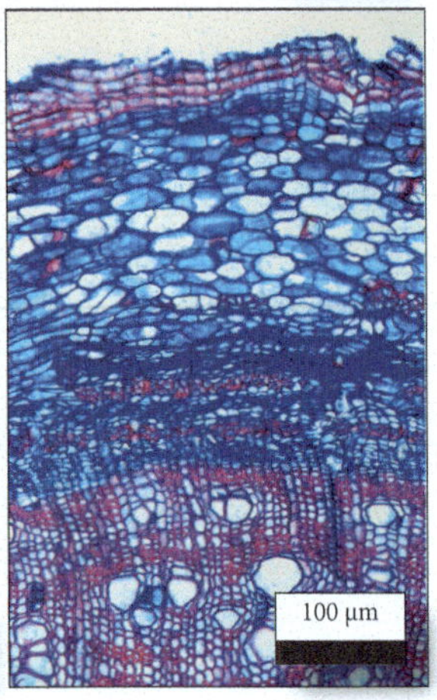

IAWA feature numbers
2 5 9 9.1 13 22 25 29 30 42 45 47 52 58 61 66 69 70 79 86 90 91 96 102.3 104 115 118 120 136 138 142 178 180 189 196

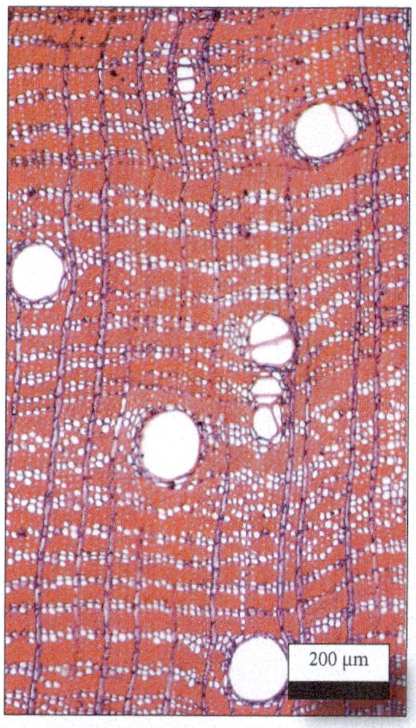

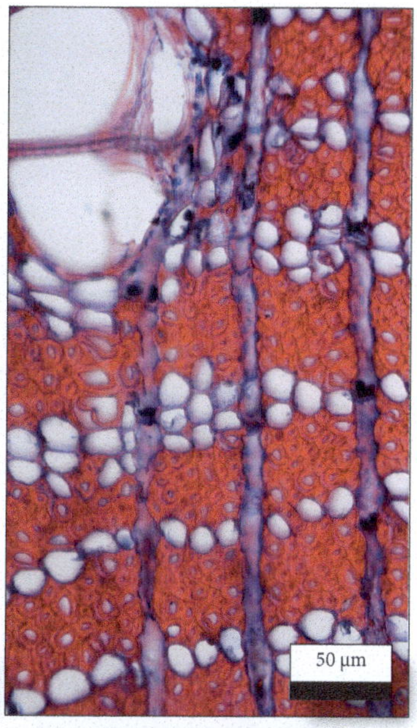

STEM XYLEM ANATOMY

Transverse section
Growth ring boundaries indistinct or absent.
Wood diffuse-porous.
Vessels solitary or in short radial multiples of 2-4 elements.
Tangential diameter of vessels 100-174-230 µm, 6-14-20 vessels/mm^2.
Vessels of two distinct diameter classes.
Gums and/or other deposits in heartwood vessels.
Fibres thick- to very thick-walled.
Apotracheal parenchyma in narrow lines up to 3 cells. Paratracheal parenchyma vasicentric.
Rays 7-8-11 rays/mm.

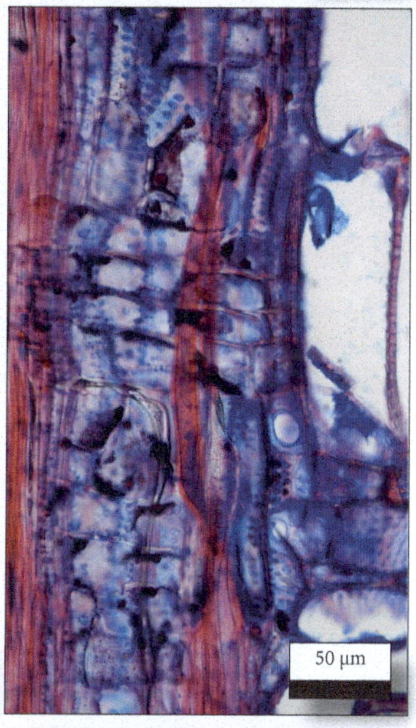

Radial section
Simple perforation plates.
Intervessel pits oval to circular and alternate, 4-7 µm in diameter.
Vestured pits present.
Vessel-ray pits with distinct borders; similar to intervessel pits in size and shape throughout the ray cell.
Vessel element length 190-295-360 µm.
Fibres with simple to minutely bordered pits.
Homogeneous rays with all cells procumbent.
Prismatic crystals in procumbent ray cells and in chambered axial parenchyma cells.

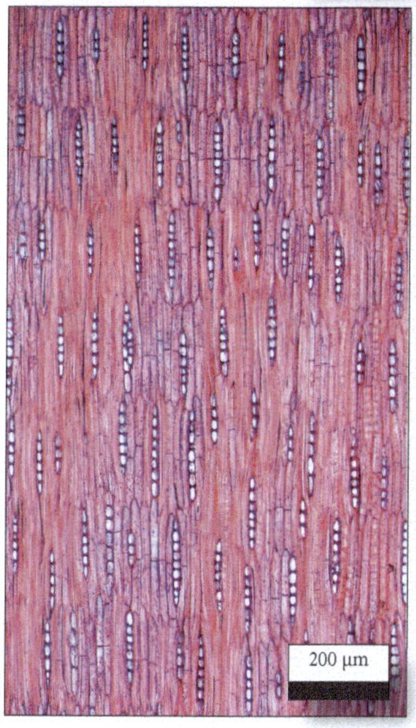

Tangential section
Rays exclusively uniseriate.
Ray height 180-240-470 µm.
Fusiform parenchyma cells present. Two cells per parenchyma strand.
Rays, vessels and axial parenchyma cells storied.
Number of ray tiers 8-9/mm.

Dalbergia maritima R.Vig.

EN: Rosewood FR: Bois de rose ML: Andramena, Bolabola

PLANT DESCRIPTION

Tree | 4-15 m high | 8-30 cm DBH
Endemic to Madagascar. Exploited species.

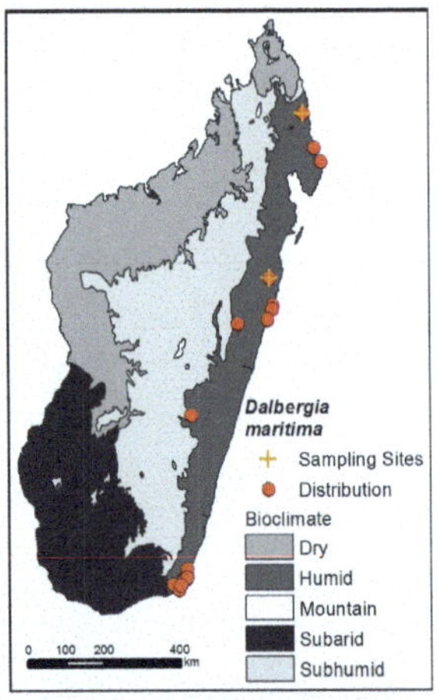

DISTRIBUTION

Elevation: 0-500 m a.s.l.
Vegetation formation: Forest
Bioclimate: Humid
Protected areas: Betampona
New protected areas: Makirovana

Slides made from material collected in Betampona.

PITH AND PRIMARY XYLEM ANATOMY

Pith: Shape of pith round. Cells dimorphic. Prismatic crystals present. Pith not visible in polarized light.

Primary xylem: Vascular bundles clearly separated. Vessels solitary and in short radial multiples.

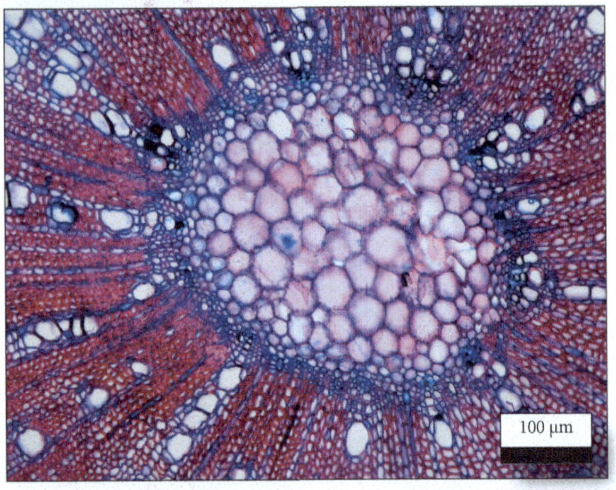

BRANCH XYLEM ANATOMY

Description: Growth rings boundaries indistinct or absent. Tangential diameter of vessels 160-113-330 µm, 8-16-22 vessels/mm^2. Vessel element length 150-183-220 µm. Intervessel pits small, 4-7 µm in diameter. Fibres thin- to thick-walled. Axial parenchyma in narrow bands up to 3 cells. Rays exclusively uniseriate, ray height 150-183-190 µm. Rays 9-11-16/mm.

Anatomical differences to stem xylem: Mean tangential diameter of vessel lumina is smaller in the branch (160-113-330/70-97-130 µm). Vessel density is higher in the branch (8-16-22/2-7-19 vessels/mm^2). Rays in the branch are slightly higher (50-183-190/60-125-180 µm) while the number of ray is similar in the branch and the stem (9-11-16/8-11-13 rays/mm).

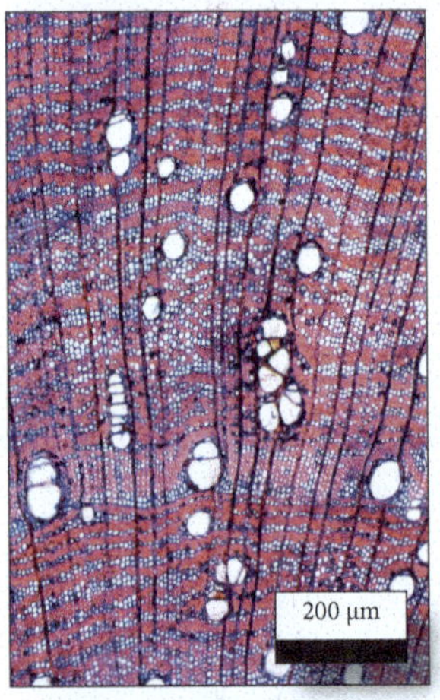

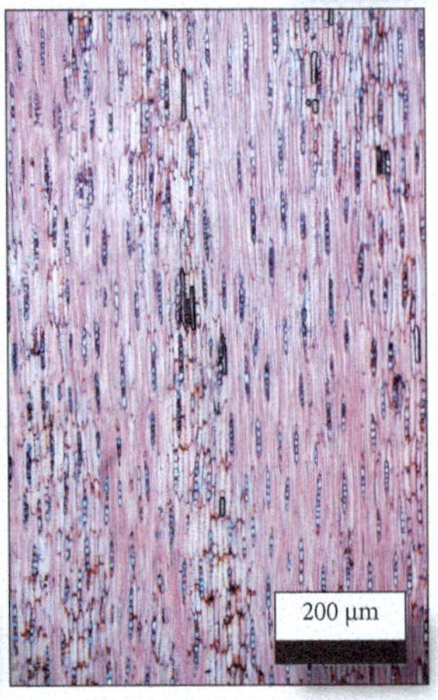

BRANCH XYLEM ANATOMY

Branch bark: Collapsed and non-collapsed sieve tubes with groups of fibres present in the phloem. The cortex consists of oval and rectangular blue-stained parenchyma cells.

Stem bark: The phloem consists of alternating bands of fibres, and collapsed and non-collapsed sieve tubes. Ray and parenchyma dilatations present. Prismatic crystals irregularly dispersed in the phloem. The phellem is homogeneous, it consists of regularly arranged rectangular cells. Lignified cells in phellem, distinct in polarized light.

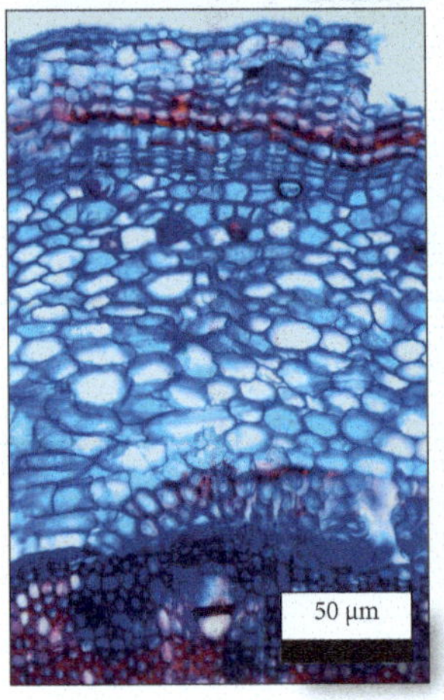

IAWA feature numbers
2 5 9 13 22 25 29 30 42 46 52 58 61 66 70 77 79 86 90
91 96 102.2 106 115 118 120 136 142 178 180 189 196

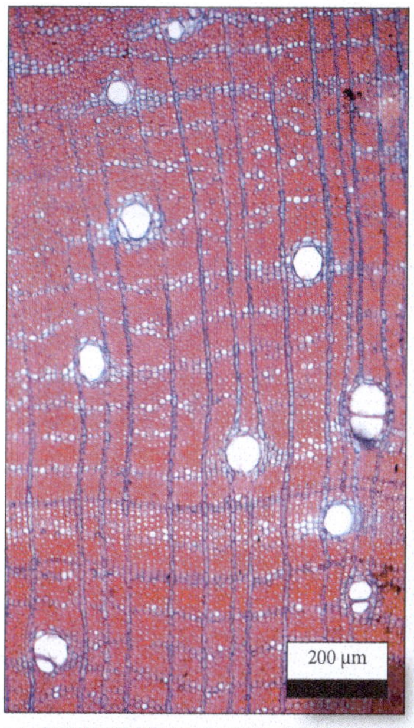

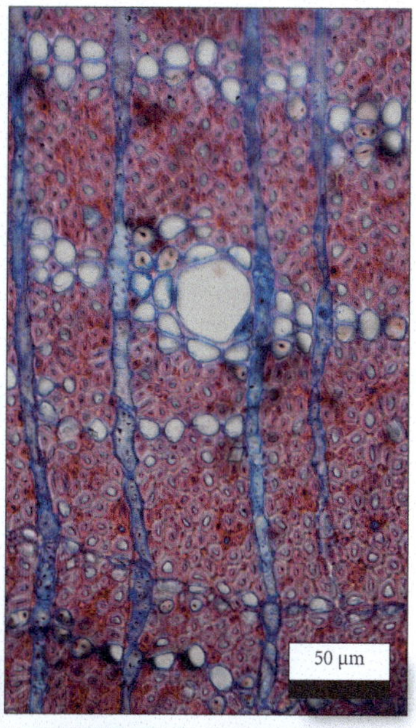

STEM XYLEM ANATOMY

Transverse section

Growth ring boundaries indistinct or absent.
Wood diffuse-porous.
Vessels predominantly solitary.
Tangential diameter of vessels 70-97-130 μm, 2-7-19 vessels/mm^2.
Gums and/or other deposits in heartwood vessels.
Fibres very thick-walled.
Apotracheal parenchyma diffuse-in-aggregates, and in narrow lines up to tree cells wide.
Paratracheal parenchyma vasicentric.
Rays 8-11-13/mm.

Radial section

Simple perforation plates.
Intervessel pits oval to circular and alternate, 4-7 μm in diameter.
Vestured pits present.
Vessel-ray pits with distinct borders; similar to intervessel pits in size and shape throughout the ray cell.
Vessel element length 150-193-240 μm.
Fibres with simple to minutely bordered pits.
Heterogenous ray with body ray cells procumbent with one row of upright and square marginal cells.
Prismatic crystal in chambered axial parenchyma cells.

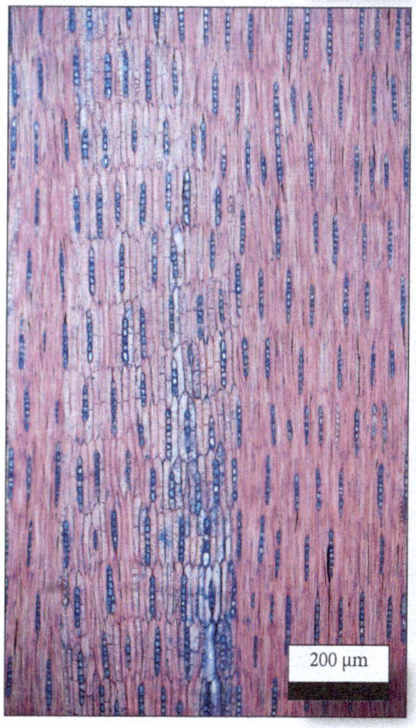

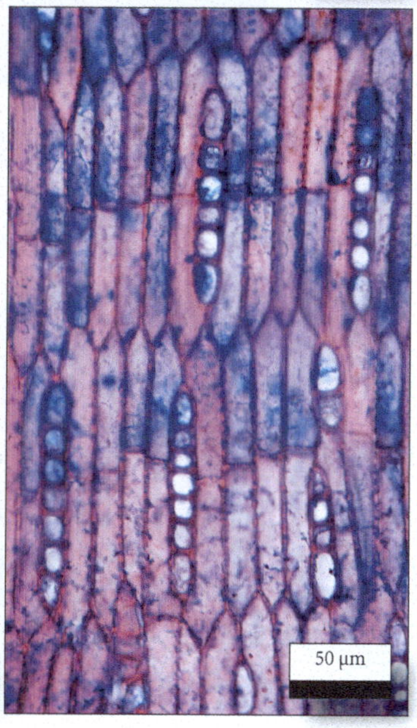

Tangential section

Rays exclusively uniseriate.
Ray height 60-125-180 μm.
Fusiform parenchyma cells present. Two cells per parenchyma strand.
Rays, vessels and axial parenchyma storied. Number of ray tiers per axial mm: 12-13.

Dalbergia mollis Bosser and R. Rabev.

Synonymy: *D. stenocarpa* var. *typica* R. Vig.

EN: Palisander FR: Palissandre ML: Manary

PLANT DESCRIPTION

Shrub to tree | 2-20 m high | up to 22 cm DBH
Endemic to Madagascar. Exploited species.

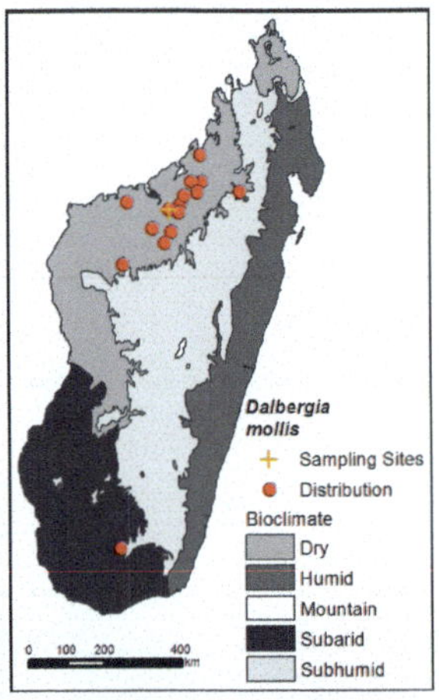

DISTRIBUTION

Elevation: 0-1000 m a.s.l.
Vegetation formation: Forest and woodland
Bioclimate: Dry, sub-arid and sub-humid
Protected areas: Ankarafantsika, Zombitsy-Vohibasia and Mikea
New protected areas: Sahafary

Slides made from material collected in Ankarafantsika.

PITH AND PRIMARY XYLEM ANATOMY

Pith: Shape of pith round to oval. Cells dimorphic. Dark staining substances rare. Simple pits in transverse cell walls. Pith not visible in polarized light.

Primary xylem: Vascular bundles not distinct. Vessels solitary and in short radial multiples.

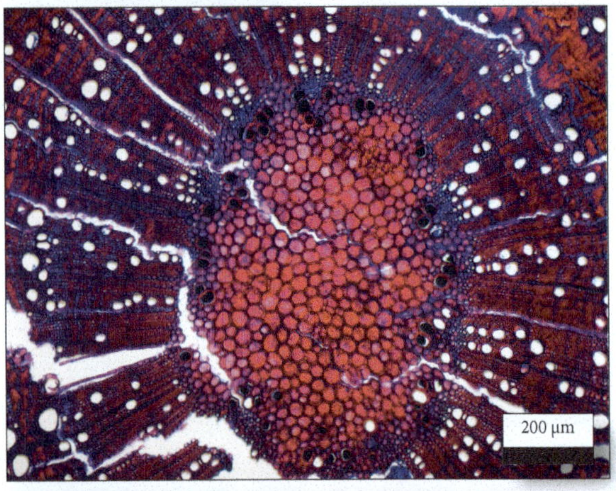

BRANCH XYLEM ANATOMY

Description: Growth rings boundaries indistinct or absent. Tangential diameter of vessels medium 40-80-150 μm, 13-23-33 vessels/mm^2. Vessel element length 70-130-220 μm. Intervessel pits small, 4-7 μm in diameter. Fibres thin- to thick-walled. Axial parenchyma diffuse-in-aggregates and in narrow lines up to 3 cells. Rays predominantly uniseriate. Ray height 50-101-150 μm, rays 8-10-13/mm.

Anatomical differences to stem xylem: Mean tangential diameter of vessel lumina in smaller in branch (40-80-150/110-169-260 μm), vessel density higher in branch (13-23-33/6-11-16 vessels/mm^2). Rays in the branch are shorter (50-101-150/60-168-420 μm) while ray number in branch and stem is similar (8-10-13/8-11-13 rays/mm).

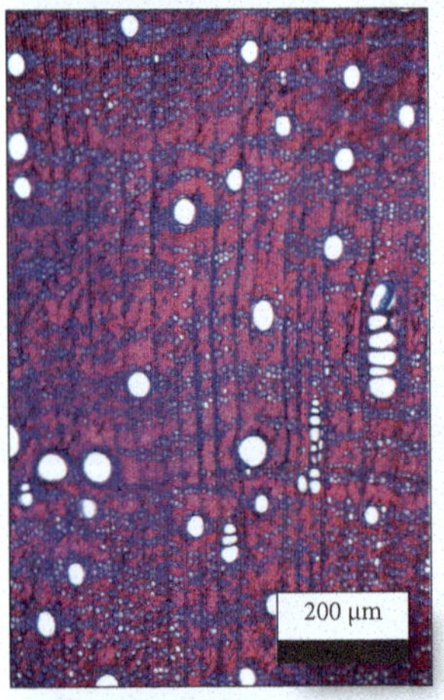

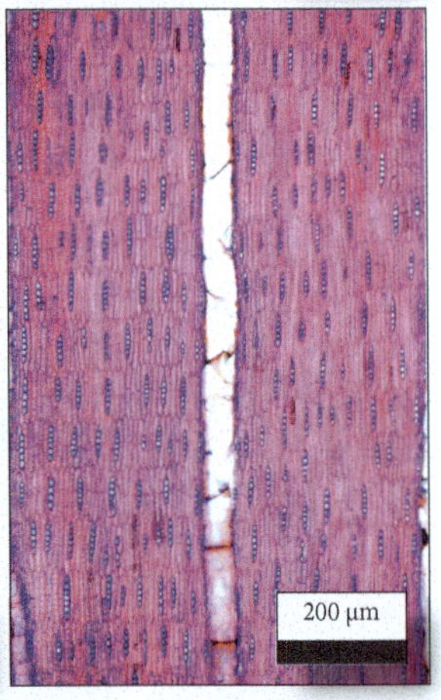

BARK ANATOMY

Branch bark: Sieve tubes, collapsed sieve tubes and fibers in groups present in the phloem. The cortex consists of irregularly shaped blue-stained parenchyma cells.

Stem bark: The phloem consists of alternating tangential bands of fibres and collapsed and non-collapsed sieve tubes. Groups of sclereids scattered or irregularly dispersed and parenchyma dilatation present in the cortex. Prismatic crystals present in the sclereids. The phellem is homogeneous and consists of regularly arranged rectangular parenchyma cells. Lignified cells in phellem, distinct in polarized light.

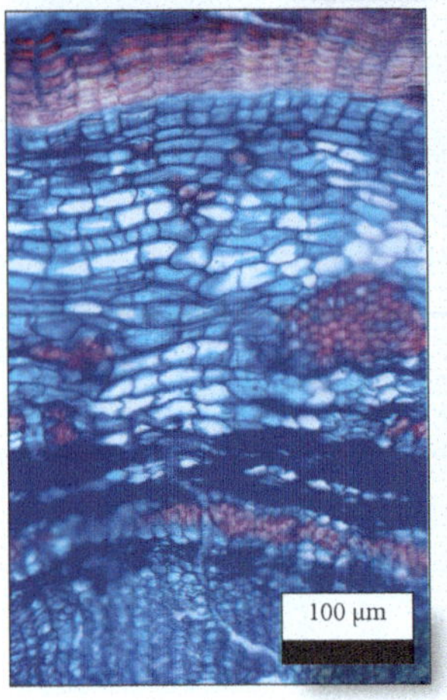

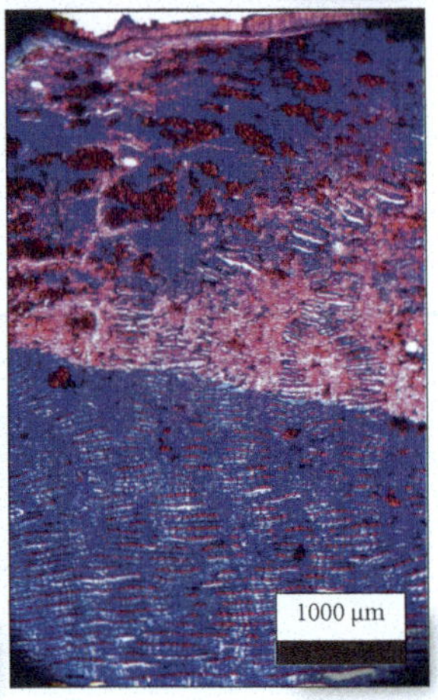

IAWA feature numbers
2 5 9 13 22 25 29 30 42 45 47 52 58 61 66 69 70 77 79 86 90 91 97 102.2 104 106 115 118 120 136 142 178 180 189 196

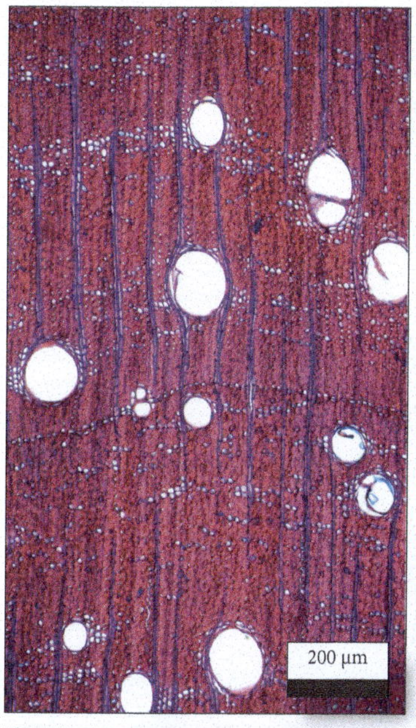

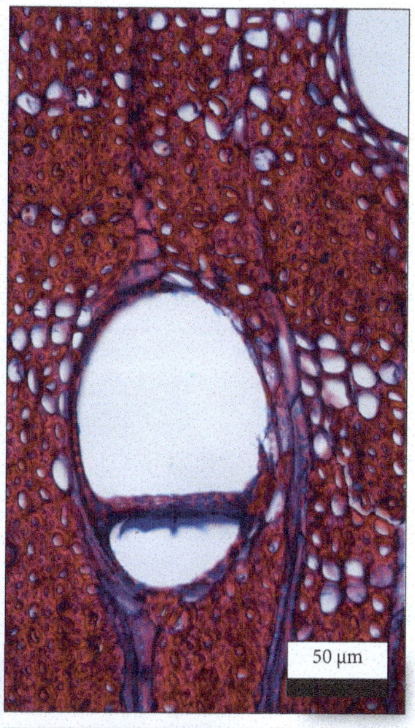

STEM XYLEM ANATOMY

Transverse section

Growth ring boundaries indistinct or absent.
Wood diffuse-porous.
Vessels predominantly solitary.
Tangential diameter of vessels 110-169-260 µm, 6-11-16 vessels/mm^2.
Vessels of two distinct diameter classes.
Gums and/or other deposits in heartwood vessels.
Fibres thick- to very thick-walled.
Apotracheal parenchyma diffuse-in-aggregates and in narrow lines up to three cells wide.
Paratracheal parenchyma vasicentric.
Rays 8-11-13/mm.

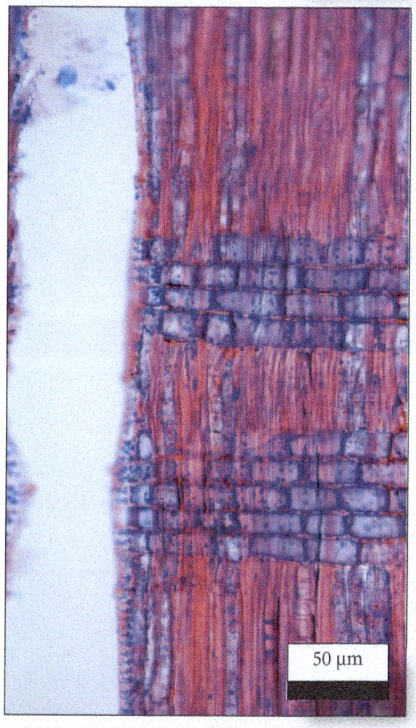

Radial section

Simple perforation plates.
Intervessel pits oval to circular and alternate, 4-7 µm in diameter.
Vestured pits present.
Vessel-ray pits with distinct borders; similar to intervessel pits in size and shape throughout the ray cell.
Vessel element length 100-156-200 µm.
Fibres with simple to minutely bordered pits.
Homogeneous and heterogeneous rays both present. In some rays body ray cells procumbent with one row of spare marginal cells.
Prismatic crystal in chambered axial parenchyma cells.

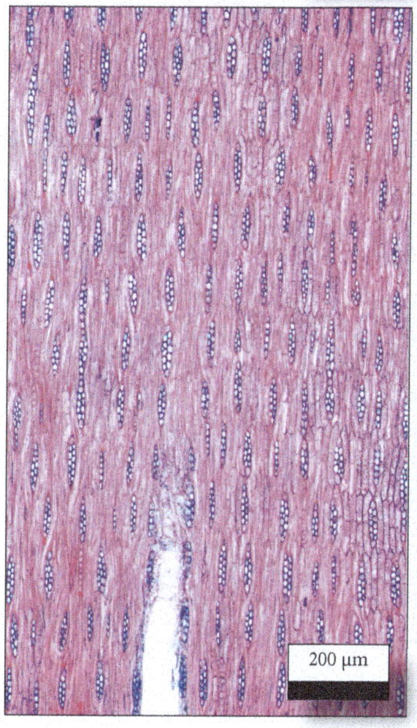

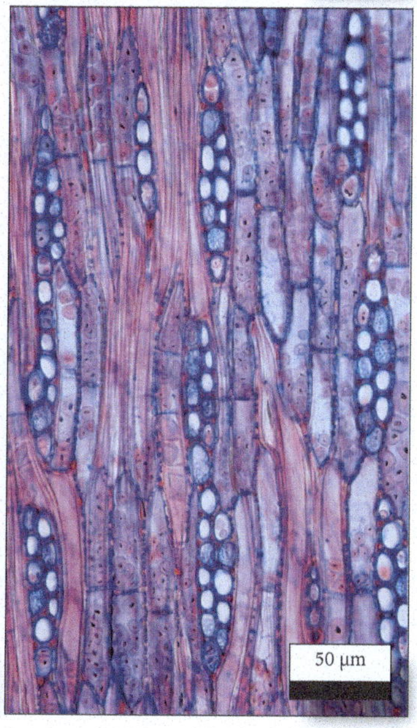

Tangential section

Rays predominantly biseriate.
Ray height 60-168-420 µm.
Fusiform parenchyma cells present. Two cells per parenchyma strand.
Rays, vessels and axial parenchyma irregularly storied. Number of ray tiers per axial mm: 12-14.

Dalbergia monticola Bosser & R. Rabev.

EN: Rosewood FR: Bois de rose ML: Voamboana mena

PLANT DESCRIPTION

Tree | 7-30 m high | 20-70 cm DBH
Endemic to Madagascar. Exploited species.

DISTRIBUTION

Elevation: 0-2000 m a.s.l.
Vegetation formation: Forest
Bioclimate: Humid and sub-humid
Protected areas: Ambohitantely, Analamazaotra-Périnet, Manongarivo, Ranomafana and Zahamena
New protected areas: Ambatovy and Didy

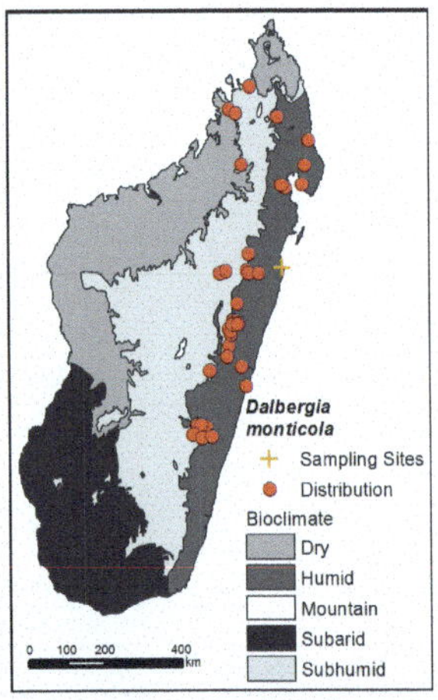

Slides made from material collected in Didy.

PITH AND PRIMARY XYLEM ANATOMY

Pith: Shape of pith round. Layer of blue-stained cells surrounds the pith. Thick-walled parenchyma cells present. Cells dimorphic. Simple pits in transverse cell walls. Pith not visible in polarized light.

Primary xylem: Vascular bundles not distinct. Vessels solitary and in radial multiples.

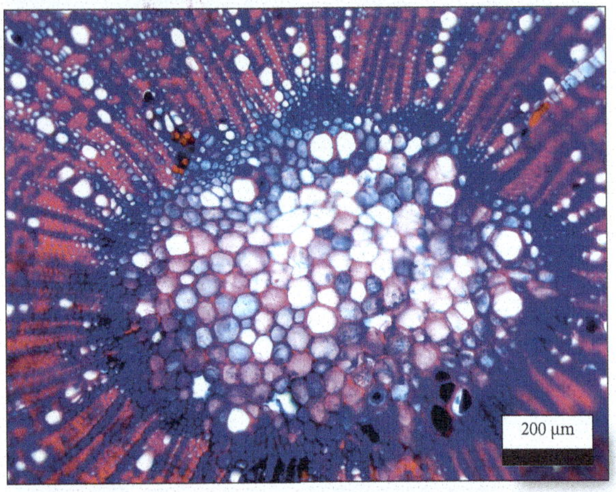

BRANCH XYLEM ANATOMY

Description: Growth rings boundaries indistinct or absent. Tangential diameter of vessels 30-43-60 µm, 6-10-17 vessels/mm². Vessel element length 180-285-410 µm. Intervessel pits small, 4-7 µm in diameter. Fibres very thick-walled. Axial parenchyma in narrow lines up to 3 cells. Rays predominantly biseriate. Ray height 70-116-180 µm, 8-10-13/mm.

Anatomical differences to stem xylem: Mean tangential diameter of vessel lumina is smaller in branch (30-43-60/130-189-290 µm) where vessel density is slightly higher (6-10-17/3-6-10 vessels per mm²). Rays in the branch are shorter (70-116-180/70-151-200 µm). Ray number is slightly lower in branch (8-10-13/9-13-20 rays/mm).

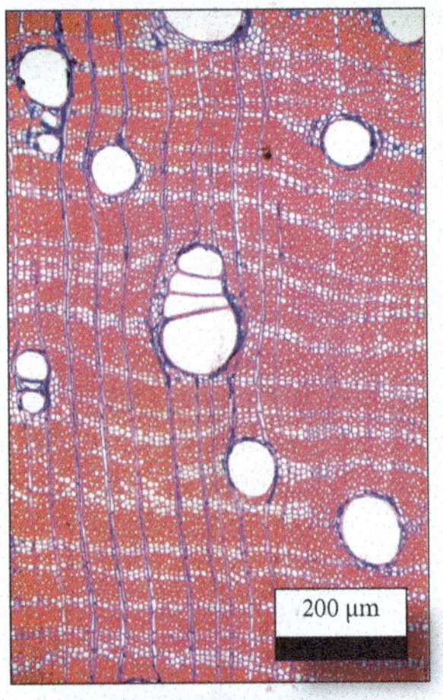

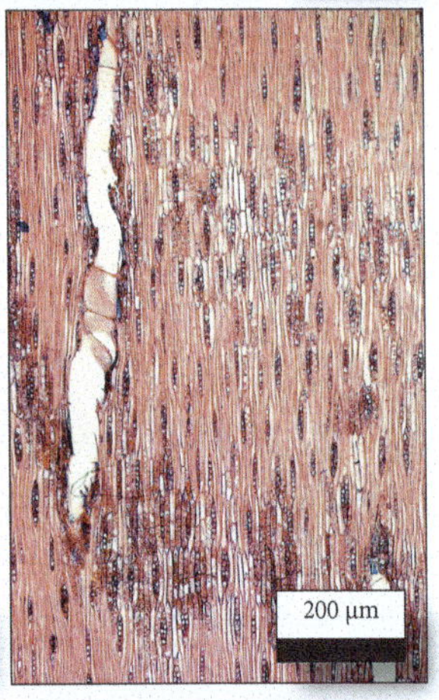

BARK ANATOMY

Branch bark: Sieve tubes and grouped fibres in tangential bands. The cortex is characterized by irregularly blue-stained parenchyma cells and irregularly dispersed groups of fibres.

Stem bark: The phloem consists of alternating bands of fibres and collapsed and non-collapsed sieve tubes. Parenchyma dilatation and prismatic crystals present. Phellem homogeneous, it consists of regularly arranged rectangular cells. Lignified cells in phellem, distinct in polarized light.

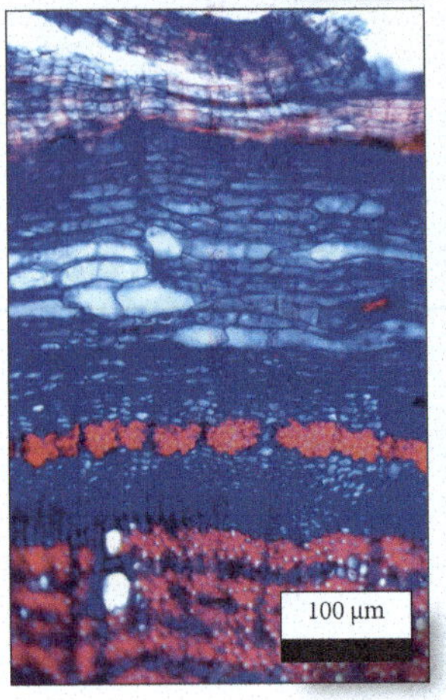

IAWA feature numbers
2 5 9 13 22 26 29 30 42 45 47 52 61 66 69 70 77 80 83
85 90 91 97 106 115 118 120 136 142 180 189 196

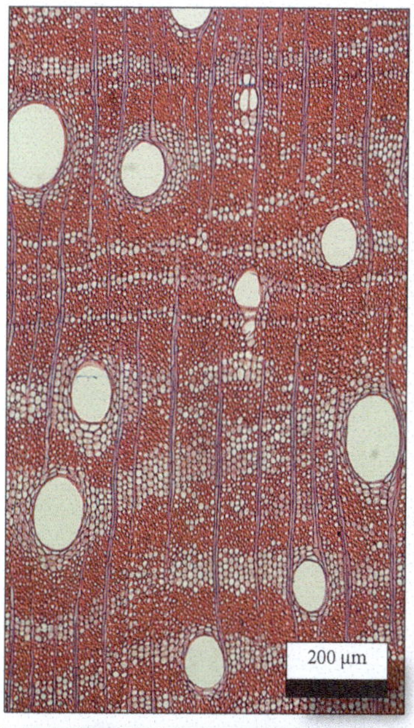

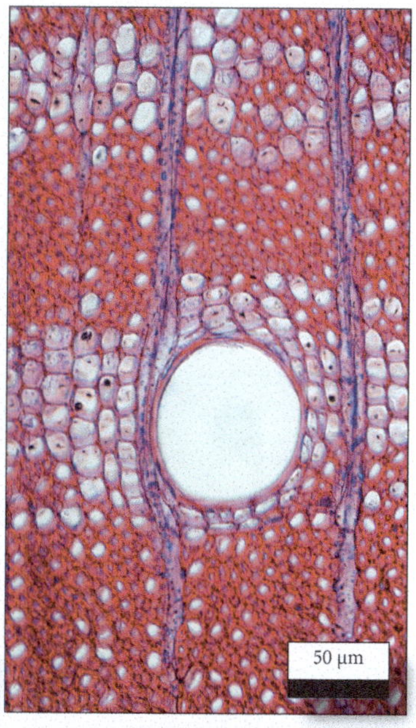

STEM XYLEM ANATOMY

Transverse section

Growth ring boundaries indistinct or absent.
Wood diffuse-porous.
Vessels mostly solitary.
Tangential diameter of vessels large 130-189-290 µm, 3-6-10 vessels per mm².
Vessels in two distinct diameter classes.
Fibers thick- to very thick-walled.
Apotracheal parenchyma diffuse-in-aggregate and in bands more than three cells wide.
Paratracheal parenchyma aliform to confluent.
Rays 9-13-20 per mm.

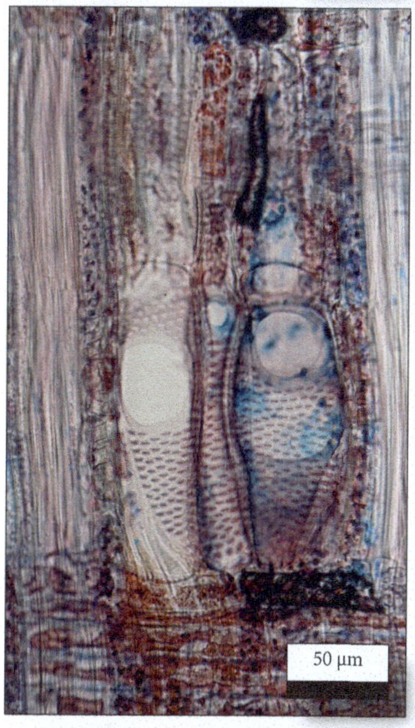

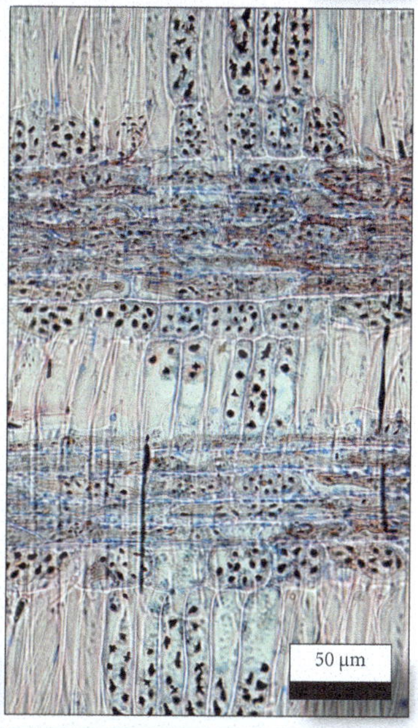

Radial section

Simple perforation plates.
Intervessel pits oval to circular and alternate, 7-10 µm in diameter.
Vestured pits present.
Vessel-ray pits with distinct borders; similar to intervessel pits in size and shape throughout the ray cell.
Vessel element length 180-194-410 µm.
Fibres with simple to minutely bordered pits.
Heterogeneous rays with body ray cells procumbent and one row of upright and square marginal cells.
Prismatic crystal in chambered axial parenchyma cells.

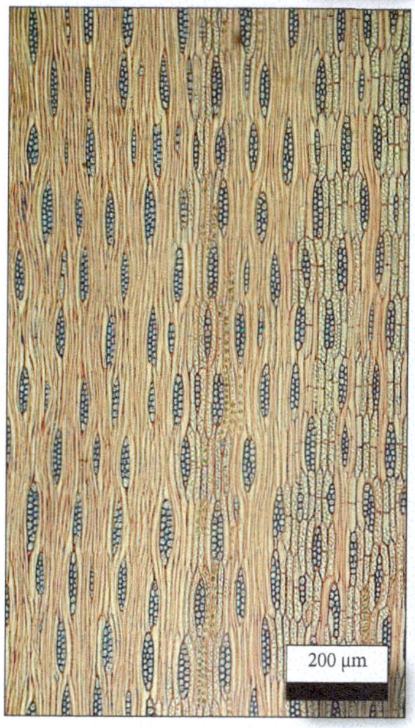

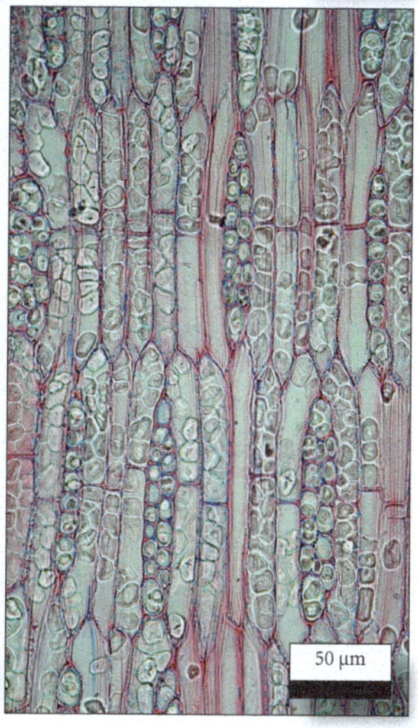

Tangential section

Rays predominantly biseriate.
Ray height 70-151-200 µm.
Fusiform parenchyma cells present. Two cells per parenchyma strand.
Rays, vessels and axial parenchyma irregularly storied. Number of ray tiers per axial mm: 12-13.

Dalbergia neoperrieri Bosser and R. Rabev.

EN: Palisander FR: Palisandre ML: Manary

PLANT DESCRIPTION

Shrub to tree | 2-15 m high
Endemic to Madagascar. Non-exploited species.

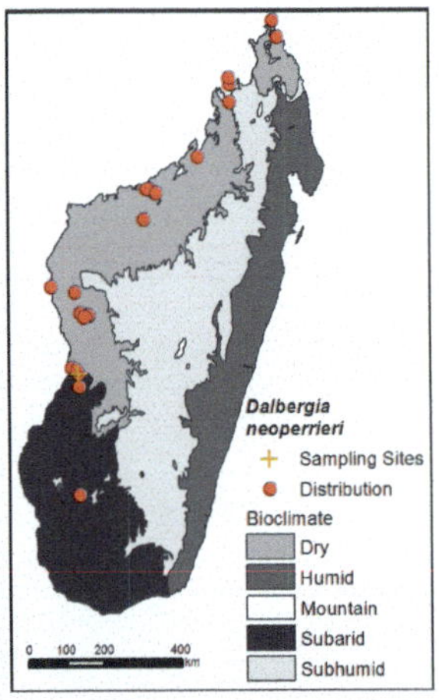

DISTRIBUTION

Elevation: 0-1000 m a.s.l.
Vegetation formation: Forest, inselberg/rock face and woodland
Bioclimate: Dry, humid and sub-humid
Protected areas: Ankarafantsika, Isalo and Kirindy-Menabe
New protected areas: Analavelona sacred forest

Slides made from material collected in Kirindy-Menabe.

PITH AND PRIMARY XYLEM ANATOMY

Pith: Shape of pith oval. Thick-walled parenchyma cells present. Cells dimorphic. Dark staining substances present. Prismatic crystals present. Simple pits in transverse cell walls. Pith not visible in polarized light.

Primary xylem: Vascular bundles not distinct. Vessels solitary and in short radial multiples. Fibers thick-walled in radial rows.

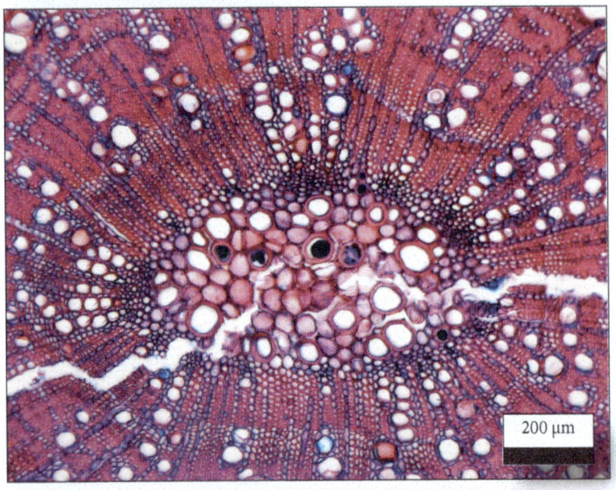

BRANCH XYLEM ANATOMY

Description: Growth ring boundaries distinct. Wood ring-porous. Tangential diameter of earlywood vessels 70-163-250 µm and latewood vessels 40-72-110 µm, 22-32-46 vessels/mm². Vessel element length 50-126-200 µm. Intervessel pits small, 4-7 µm in diameter. Fibres thick- to very thick-walled. Axial parenchyma diffuse-in-aggregates, and in narrow lines. Rays predominantly uniseriate. Ray height 30-103-210 µm, rays 7-10-13 per mm.

Anatomical differences to stem xylem: Earlywood vessels size large comparable (70-163-250/80-175-280 µm), latewood vessels smaller in branch (40-53-80/40-70-110 µm). Vessel number is similar in stem and branch (22-32-46/21-35-45 vessels/mm²). Ray height and number are also similar (30-103-210/60-110-230 µm; 7-10-13/9-11-13 rays/mm).

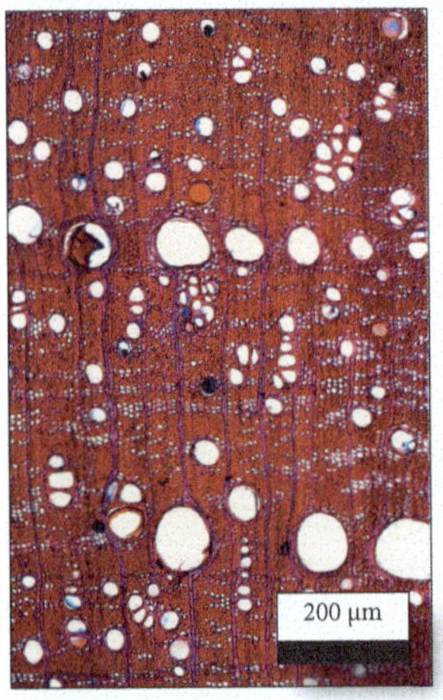

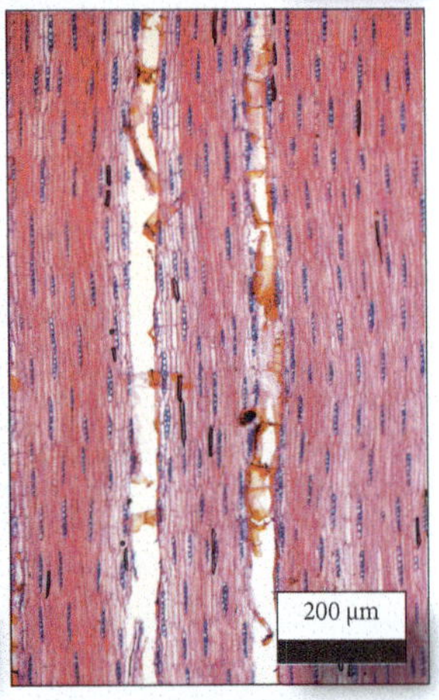

BARK ANATOMY

Branch bark: Sieve tubes arrangement in tangential row. Collapsed sieve tubes alternate with groups of thick-walled fibres in tangential bands. The cortex consists of square and rectangular blue-stained parenchyma cells. Prismatic crystal scattered or irregularly dispersed in the phloem and cortex.

Stem bark: The phloem consists of alternating tangential bands of fibres and collapsed and non-collapsed sieve tubes. Axial parenchyma dilatation present. The phellem is homogeneous, it consists of regularly arranged rectangular cells. Lignified cells in phellem, distinct in polarized light.

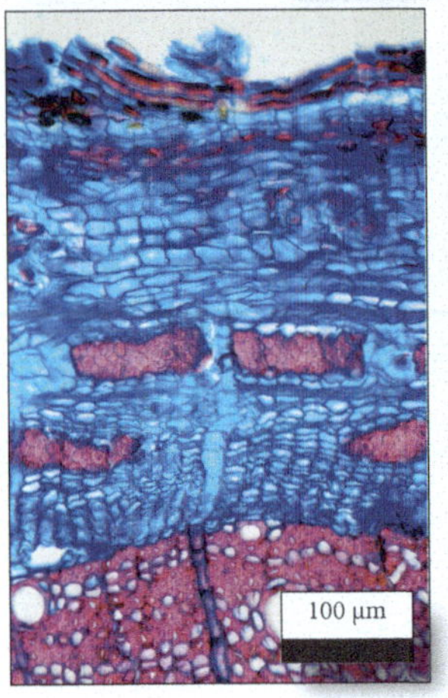

IAWA feature numbers

1 3 13 22 25 29 30 42 48 52 58 61 66 69 70 77 78 90 91
96.1 104 106 115 118 120 136 142 180 189 196

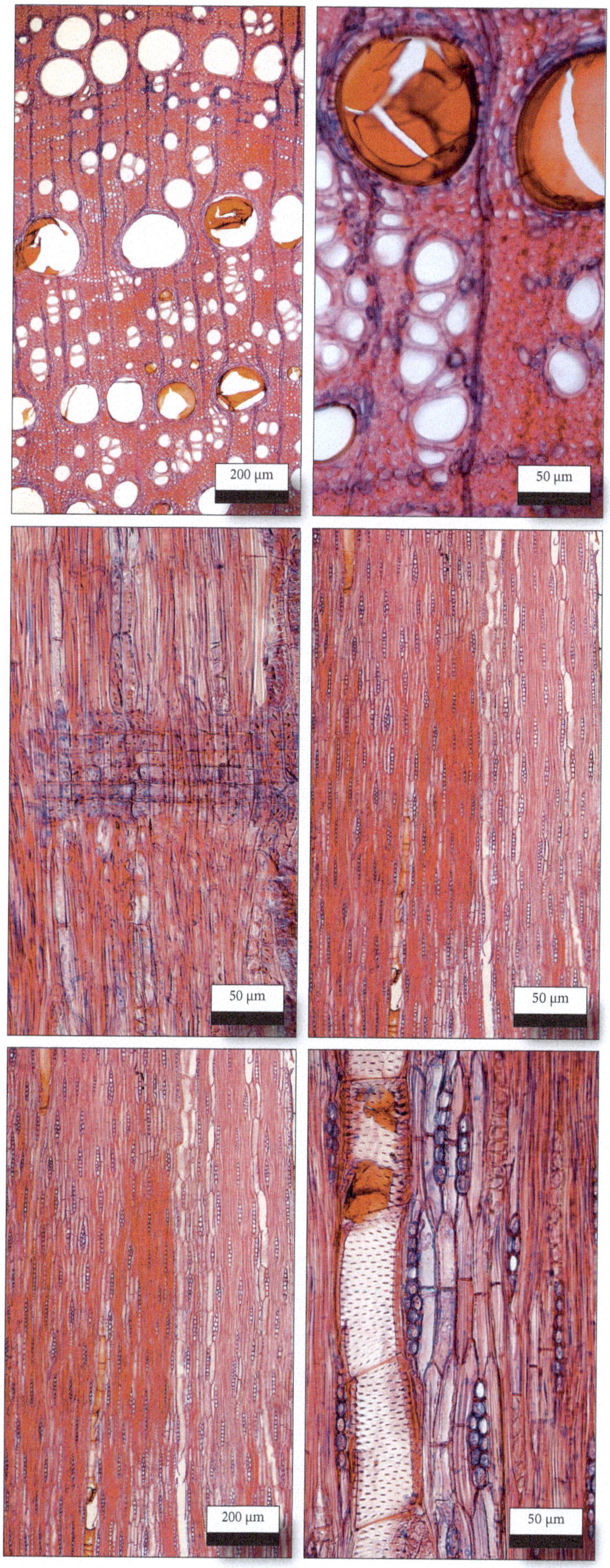

STEM XYLEM ANATOMY

Transverse section

Growth ring boundaries distinct, marked by thick-walled and radially flattened latewood fibres and by differences in vessel diameter between latewood and earlywood.
Wood ring-porous.
Earlywood vessels solitary, latewood vessels in radial multiples of 2-6 elements, or in clusters.
Tangential diameter of earlywood vessels 80-175-280 µm and of latewood vessels 40-53-80 µm, 21-35-45 vessels per mm^2.
Gums and/or other deposits in heartwood vessels.
Fibres thick- to very thick-walled.
Apotracheal parenchyma diffuse-in-aggregates.
Paratracheal parenchyma scanty.
Rays 9-11-13 per mm.

Radial section

Simple perforation plates.
Intervessel pits oval to circular and alternate, 4-7 µm in diameter.
Vestured pits present.
Vessel-ray pits with distinct borders; similar to intervessel pits in size and shape throughout the ray cell.
Vessel element length 50-126-200 µm.
Fibres with simple to minutely bordered pits.
Heterogeneous and homogeneous rays both present. Some rays with body ray cells procumbent with one row of square marginal cells.
Prismatic crystal in chambered axial parenchyma cells.

Tangential section

Rays predominantly uniseriate, a few biseriate rays occur.
Ray height 60-110-230 µm.
Fusiform parenchyma cells present. Two cells per parenchyma strand.
Rays, vessels and axial parenchyma storied.
Number of ray tiers per axial mm: 16-17.

Dalbergia orientalis Bosser and R.Rabev.

EN: Palisandre FR: Palissandre ML: Voamboana

PLANT DESCRIPTION

Tree | 7-15 m high | 25-40 cm DBH
Endemic to Madagascar. Exploited species.

DISTRIBUTION

Elevation: 0-1500 m a.s.l.
Vegetation formation: Forest
Bioclimate: Humid
Protected areas: Betampona
New protected areas: Mahabo-Mananivo

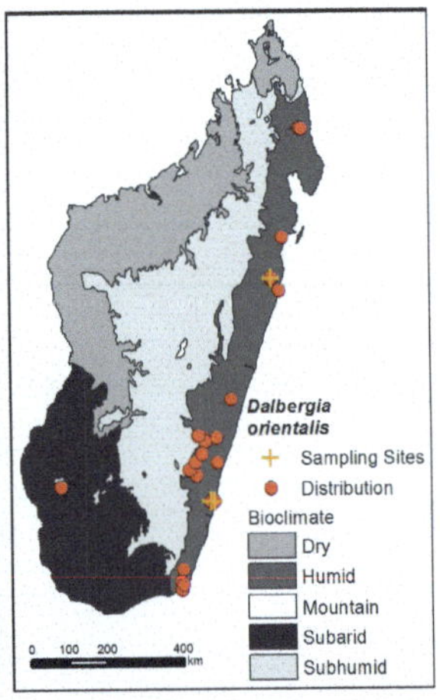

Slides made from material collected in Betampona.

PITH AND PRIMARY XYLEM ANATOMY

Pith: Shape of pith round. Thick-walled parenchyma cells present. Cells dimorphic. Prismatic crystals rare. Simple pits in transverse cell walls. Pith not visible in polarized light.

Primary xylem: Vascular bundles clearly separated. Vessels solitary and in short radial multiples. Fibres thin- to thick-walled in radial rows.

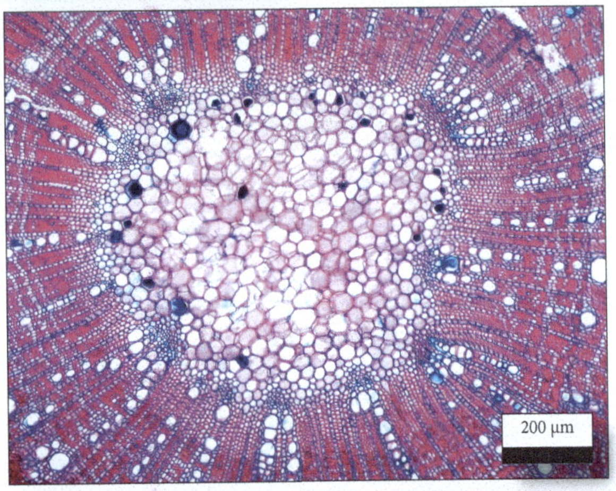

BRANCH XYLEM ANATOMY

Description: Growth rings boundaries indistinct or absent. Tangential diameter of vessels 80-126-180 µm, 23-36-49 vessels/mm². Vessel element length 80-145-180 µm. Intervessel pits small, 4-7 µm in diameter. Fibres thin- to thick-walled. Axial parenchyma in narrow bands up to 3 cells wide. Rays predominantly biseriate. Ray height 60-119-210 µm, rays 8-11-14/mm.

Anatomical differences to stem xylem: Mean tangential diameter of vessel lumina is smaller in branch (80-126-180/140-204-300 µm). Vessel number is higher in the branch (23-36-49/12-18-25 vessels/mm²). Ray height and number are higher in the branch (60-119-210/60-115-150 µm, 8-11-14/6-9-12 rays/mm).

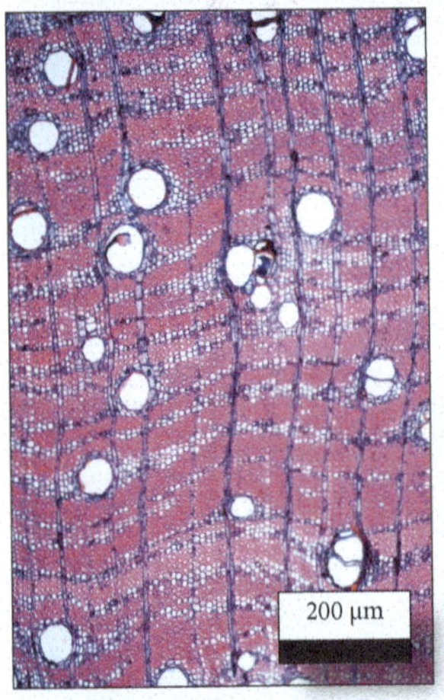

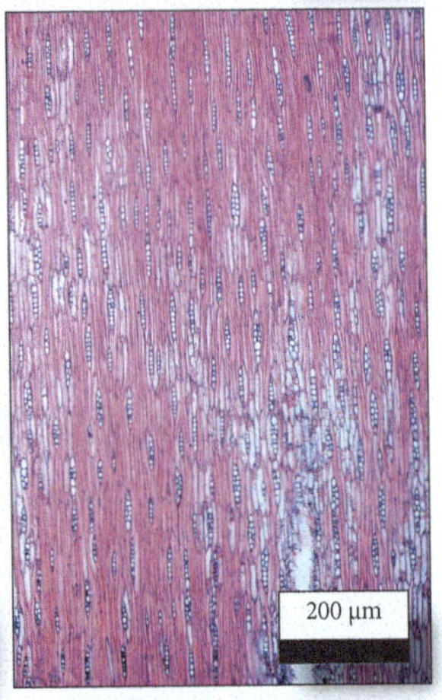

BARK ANATOMY

Branch bark: Collapsed and non-collapsed sieve tubes in the phloem. Prismatic crystals and fibres grouped tangentially present. The cortex consists of irregular round to oval and rectangular blue-stained parenchyma cells.

Stem bark: The phloem consists of alternating bands of fibres and collapsed and non-collapsed sieve tubes. Distinct ray dilatation, but only some rays become dilated. Sand crystals scattered or irregularly dispersed in phloem. The phellem is homogeneous, consists of regularly arranged rectangular cells. Lignified cells in phellem, distinct in polarized light.

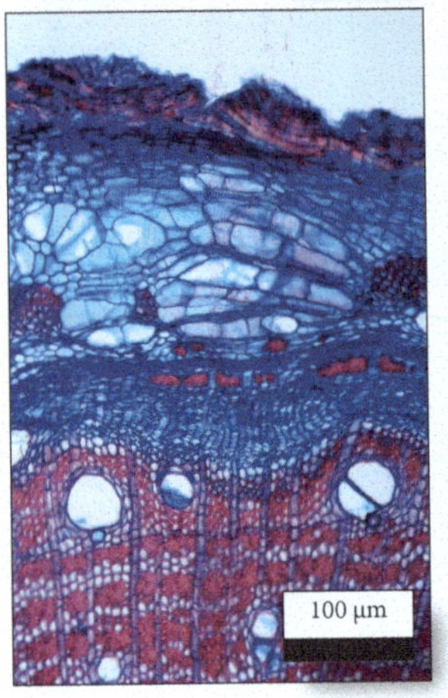

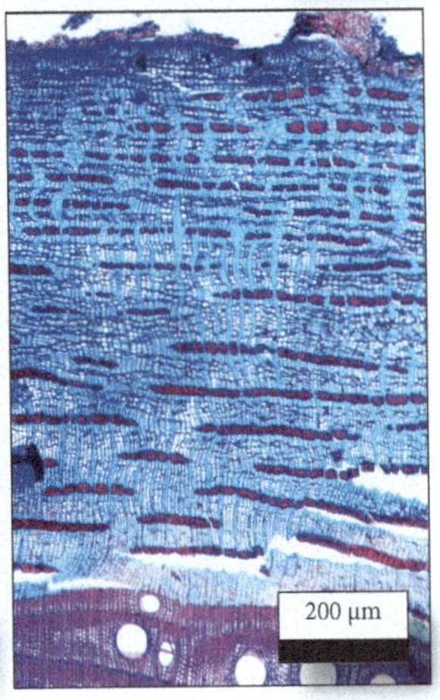

IAWA feature numbers
1 2 5 9 13 22 25 29 30 43 45 46 52 58 61 66 69 79 83 85 90 91 97 102.2 106 115 118 120 136 142 180 189 196

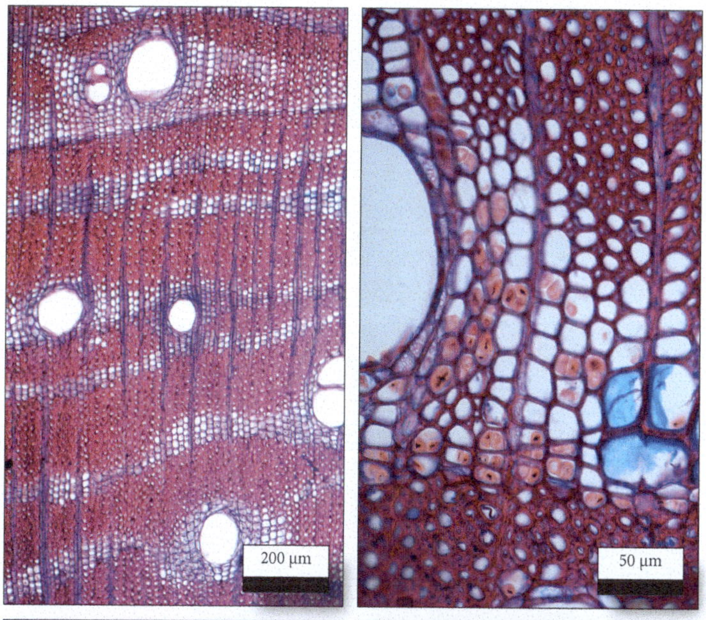

STEM XYLEM ANATOMY

Transverse section

Growth rings boundaries distinct, marked by radially flattened latewood fibres.
Wood diffuse-porous.
Vessels mostly solitary.
Tangential diameter of vessels 140-204-300 µm, 12-18-25 vessels/mm^2.
Gums and/or other deposits in heartwood vessels.
Fibres thin- to thick-walled.
Parenchyma in bands more than three cells wide. Paratracheal parenchyma vasicentric to confluent.
Rays 6-9-12/mm.

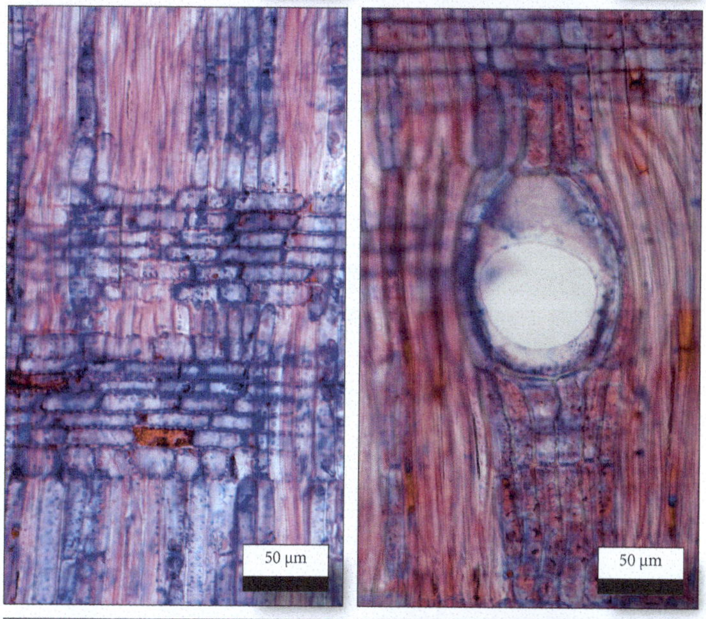

Radial section

Simple perforation plates.
Intervessel pits oval to circular and alternate, 4-7 µm in diameter.
Vestured pits present.
Vessel-ray pits with distinct borders; similar to intervessel pits in size and shape throughout the ray cell.
Vessel element length 130-169-200 µm.
Fibres with simple to minutely bordered pits.
Heterogeneous rays with body ray cells procumbent and one row of upright and square marginal cells.
Prismatic crystal in chambered axial parenchyma cells.

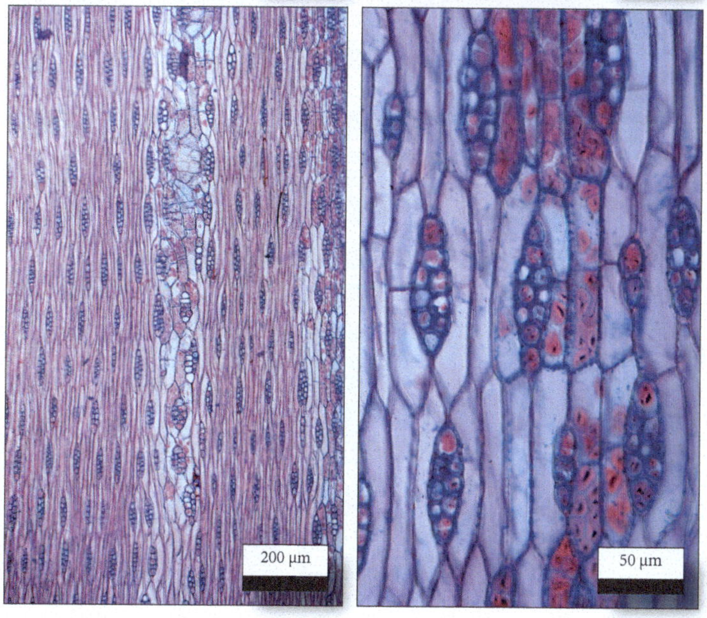

Tangential section

Rays predominantly biseriate.
Ray height 60-115-150 µm.
Fusiform parenchyma cells present. Two cells per parenchyma strand.
Rays, vessels and axial parenchyma storied.
Number of ray tiers per axial mm: 14-15.

Dalbergia peltieri Bosser and R. Rabev.

Synonymy: *D. microcarpa* R. Vig.

EN: Palisander FR: Palissandre ML: Manary

PLANT DESCRIPTION

Shrub to tree | 5-12 m high | 5-60 cm DBH
Endemic to Madagascar. Exploited species.

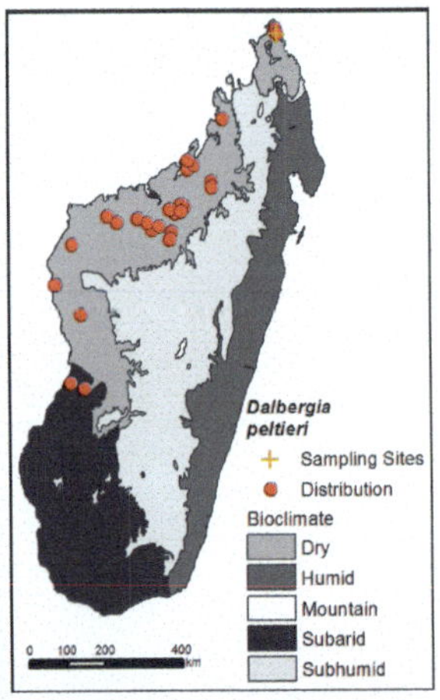

DISTRIBUTION

Elevation: 0-500 m a.s.l.
Vegetation formation: Forest and woodland
Bioclimat: Dry and sub-humid
Protected areas: Ankarafantsika, Bemaraha and Namoroka
New protected areas: Orangea-Rigny

Slides made from material collected in Orangea-Rigny.

PITH AND PRIMARY XYLEM ANATOMY

Pith: Shape of pith round. Thick-walled parenchyma cells present. Cells dimorphic. Dark staining substances present. Prismatic crystals rare. Simple pits in transverse cell walls. Pith not visible in polarized light.

Primary xylem: Vascular bundles clearly separated. Vessels solitary and in short radial multiples. Fibres thin- to thick-walled in radial rows.

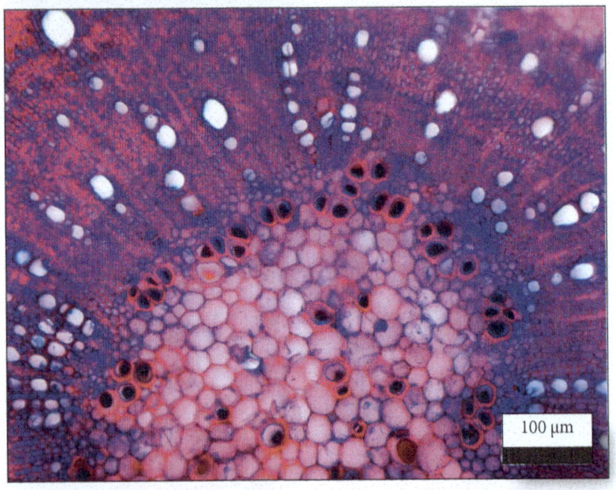

BRANCH XYLEM ANATOMY

Description: Growth ring boundaries indistinct. Tangential diameter of vessels 40-62-120 µm, 17-24-37 vessels/mm^2. Vessel element length 110-140-180 µm. Intervessel pits small, 4-7 µm in diameter. Fibres thick to very thick-walled. Axial parenchyma in bands more than 3 cells. Rays predominantly uniseriate. Ray height 70-139-230 µm, rays 9-11-13/mm.

Anatomical differences to stem xylem: Mean tangential diameter of vessel lumina is smaller in branch (40-62-120/40-88-150 µm). Vessel density is higher in the branch than in stem (17-24-37/10-16-24 vessels/mm^2). Rays are slightly higher in the branch (70-139-230/50-131-200 µm) while ray density is similar (9-11-13/10-11-14 rays/mm).

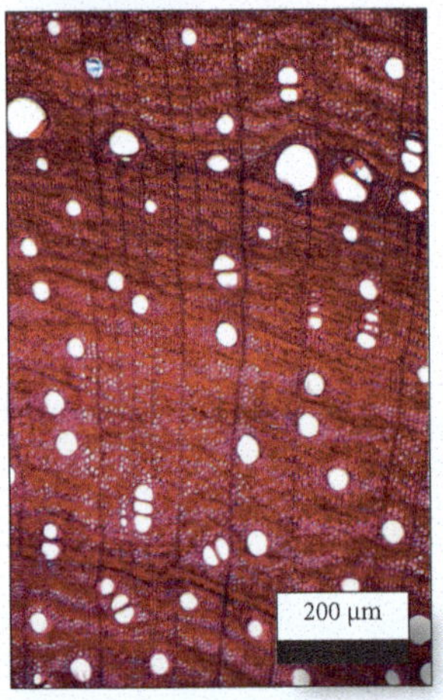

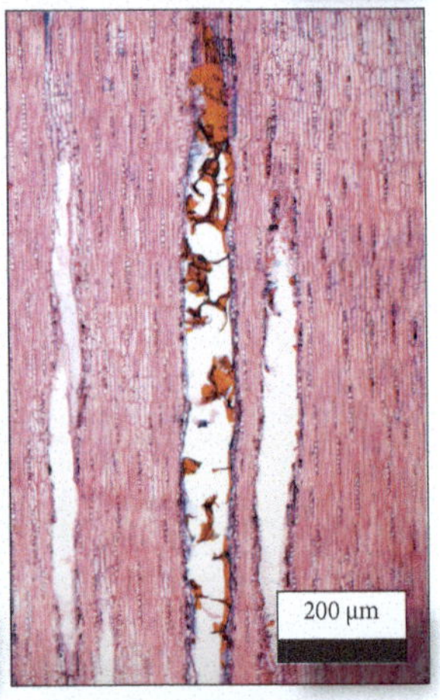

BARK ANATOMY

Branch bark: Collapsed and non-collapsed sieve tubes present in the phloem. The cortex consists of irregular rectangular blue-stained parenchyma cells and groups of tangentially distributed fibres.

Stem bark: The phloem consists of alternating bands of fibres and collapsed and non-collapsed sieve tubes. Parenchyma dilatations present. Prismatic crystals irregularly dispersed in phloem. The phellem is homogeneous, it consists of regularly arranged rectangular cells. Lignified cells in phellem distinct in polarized light.

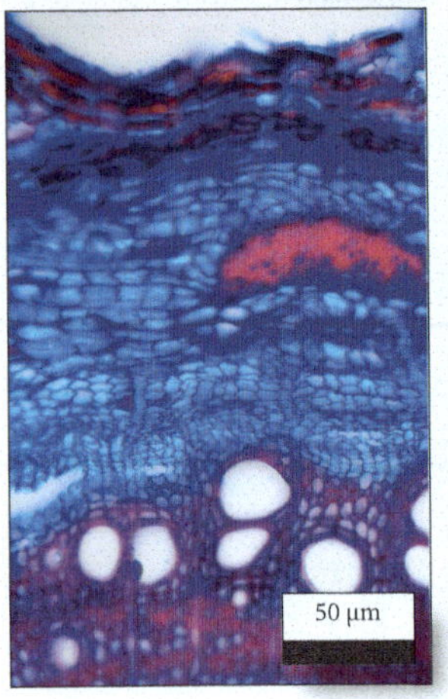

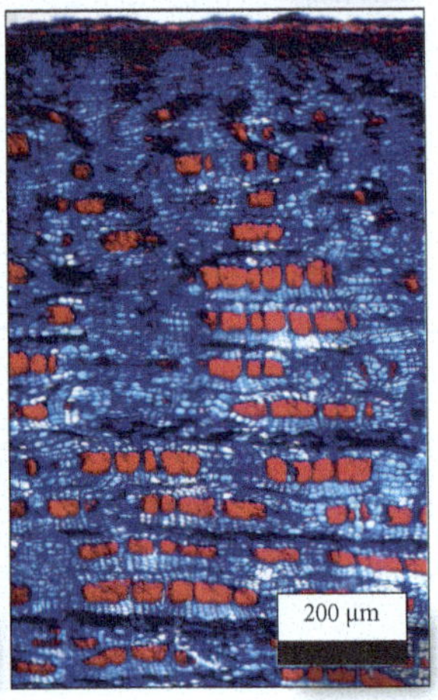

Dalbergia peltieri Bosser and R. Rabev.

IAWA feature numbers

1 3 9 11 13 22 26 29 30 42 47 52 58 61 66 69 77 79 82
86 90 91 97 102.2 104 115 118 120 136 142 180 189 196

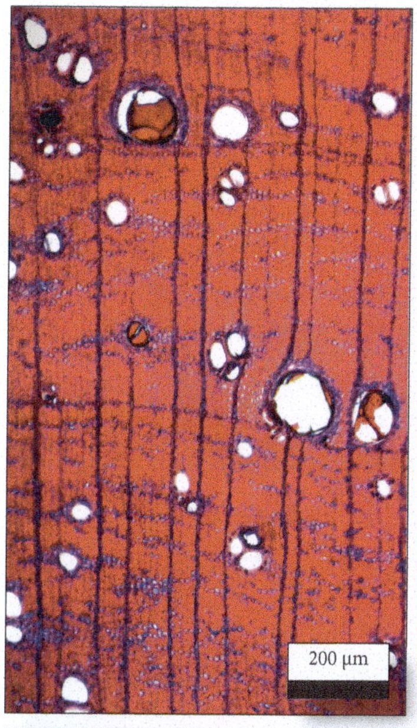

STEM XYLEM ANATOMY

Transverse section

Growth ring boundaries distinct, marked by radially flattened latewood fibres and by difference vessel diameter between latewood and earlywood.
Wood ring-porous.
Vessels solitary and in clusters of 3 to 4 elements.
Tangential diameter of earlywood vessels 90-136-190 µm and latewood vessels 40-88-150 µm, 10-16-24 vessels/mm².
Gums and/or other deposits in heartwood vessels.
Fibres thin- to thick-walled.
Apotracheal parenchyma diffuse-in-aggregates.
Paratracheal parenchyma vasicentric to winged-aliform. Axial parenchyma bands more than three cells wide.
Rays 10-11-14/mm.

Radial section

Simple perforation plates.
Intervessel pits oval to circular and alternate, 7-10 µm in diameter.
Vestured pits present.
Vessel-ray pits with distinct borders; similar to intervessel pits in size and shape throughout the ray cell.
Vessel element length 100-148-180 µm.
Fibres with simple to minutely bordered pits.
Homogeneous and heterogeneous rays both present. A few rays with body ray cells procumbent and one row of square or upright cells.
Prismatic crystal in chambered axial parenchyma cells.

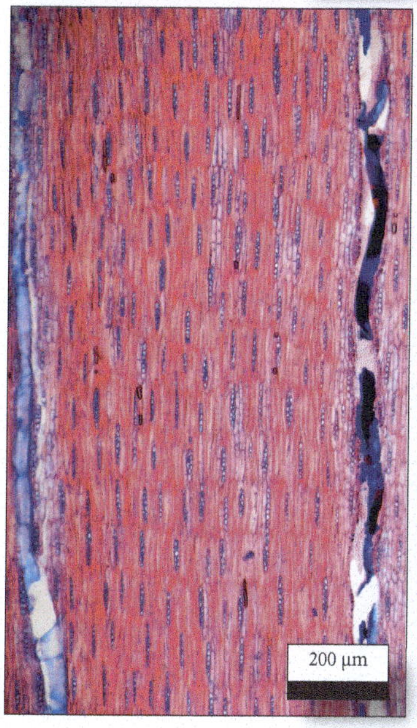

Tangential section

Rays predominantly biseriate.
Ray height 50-131-200 µm.
Fusiform parenchyma cells present. Two cells per parenchyma strand.
Rays, vessels and axial parenchyma irregularly storied. Number of ray tiers per axial mm: 16-17.

Dalbergia pervillei Vatke

Synonymy: *D. densicoma* Baill., *D. obtusa* Lecompte, *D. retusa* Baill.

EN: Palisandre FR: Palissandre ML: Voamboana sovoka

PLANT DESCRIPTION

Shrub to tree | 3-15 m high | 12-40 cm DBH
Endemic to Madagascar. Exploited species.

DISTRIBUTION

Elevation: 0-1000 m a.s.l.
Vegetation formation: Forest and woodland
Bioclimate: Dry, sub-arid and sub-humid
Protected areas: Andohahela, Ankarafantsika, Isalo, Manongarivo, Namoroka and Zombitsy-Vohibasia
New protected areas: Montagne des Français, Orangea/Rigny and Sahafary

Slides made from material collected in Montagne des Français.

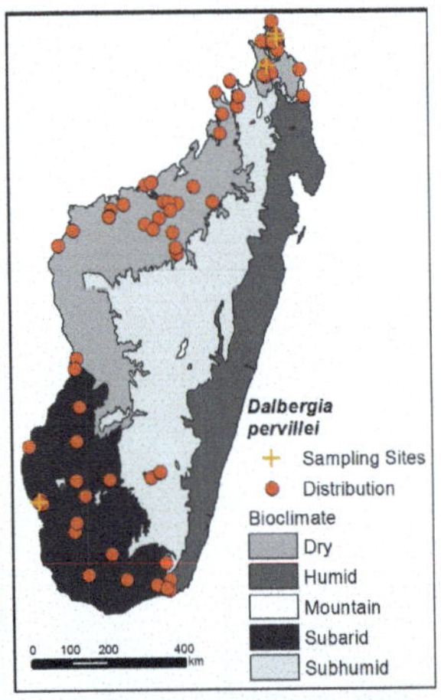

PITH AND PRIMARY XYLEM ANATOMY

Pith: Shape of pith polygonal. Layers of regularly red stained cells surround the pith. Cells dimorphic. Prismatic crystals abundant. Simple pits in transverse cell walls. Pith not visible in polarized light.

Primary xylem: Vascular bundles clearly separated. Vessels solitary and in short radial multiples.

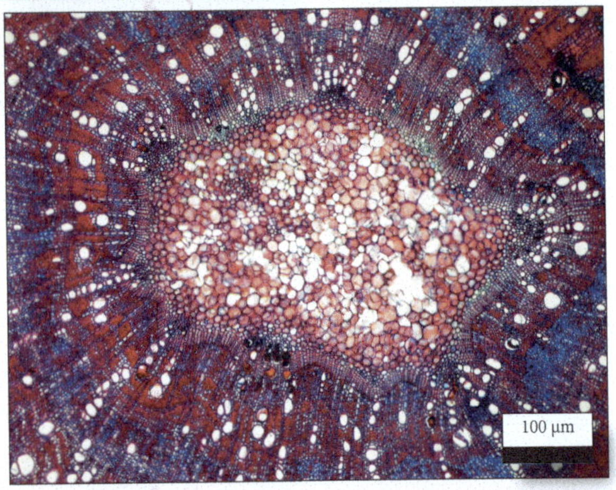

BRANCH XYLEM ANATOMY

Description: Growth rings boundaries indistinct or absent. Tangential diameter of vessels 70-89-120 µm, 13-26-45 vessels/mm^2. Vessel element length 130-172-300 µm. Intervessel pits small, 4-7 µm in diameter. Fibres thin- to thick-walled. Axial parenchyma diffuse-in-aggregates, and winged-aliform. Rays uni- and biseriate, 30-113-200 µm high. Rays 5-9-13/mm.

Anatomical differences to stem xylem: Mean tangential diameter of vessel lumina is smaller in branch (70-89-120/110-134-180 µm). Vessel density in the branch is slightly higher (13-26-45/16-20-33 vessels/mm^2). Ray height is similar (30-113-200/80-117-160 µm), while ray density is smaller in branch (5-9-13/20-21-22 rays/mm). Biseriate rays present in branch. Axial elements are not storied in the branch.

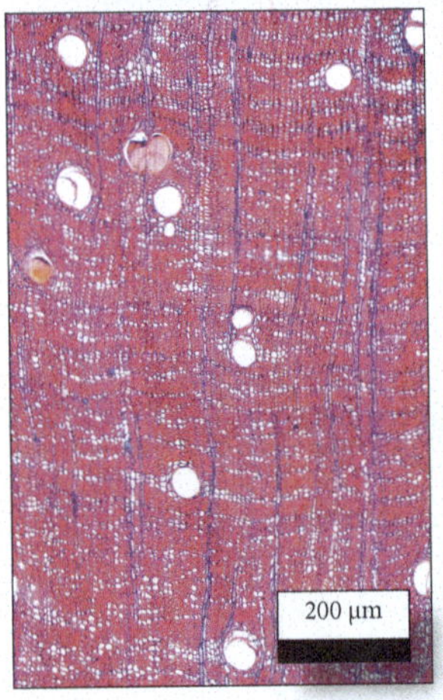

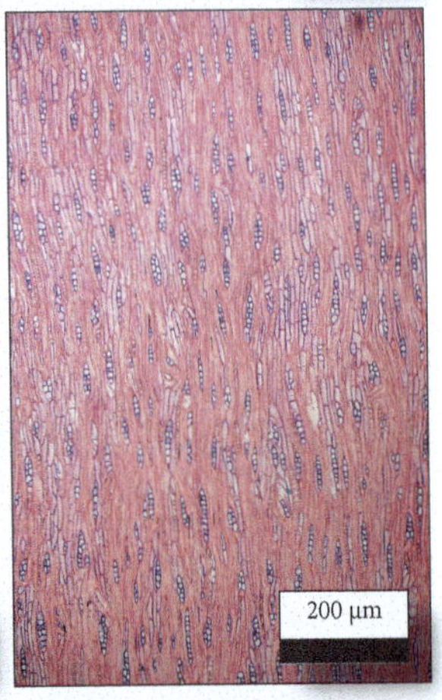

BARK ANATOMY

Branch bark: Collapsed and non-collapsed sieve tubes present in the phloem. Groups of fibres in tangential bands.
The cortex consists of irregular round and oval blue-stained parenchyma cells, and a layer of red-stained cells near the phellem.

Stem bark: The phloem consists on alternating bands of fibres and collapsed and non-collapsed sieve tubes. Crystal sand irregularly dispersed in phloem and phellem. The layered phellem is homogeneous and consists of regularly arranged rectangular parenchyma cells with dark staining substances. Lignified cells in phellem, distinct in polarized light.

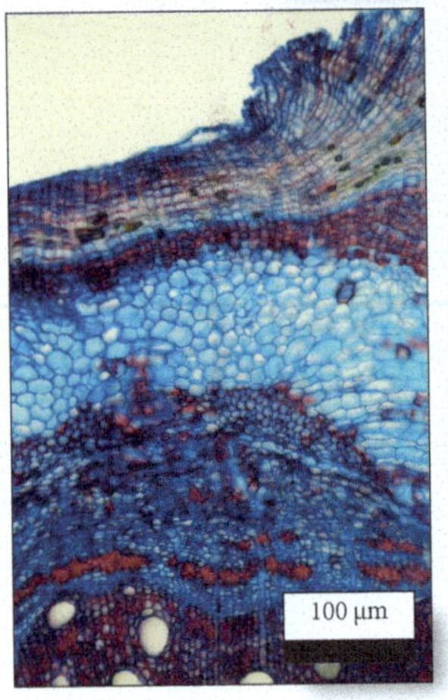

IAWA feature numbers

2 5 9 9.1 13 22 25 29 30 42 48 52 58 61 66 70 77 79 80
90 91 96 102.2 104 106 116 118 120 136 142 180 189 196

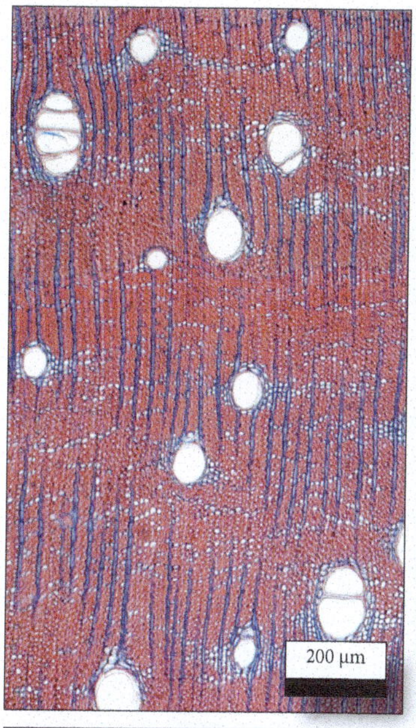

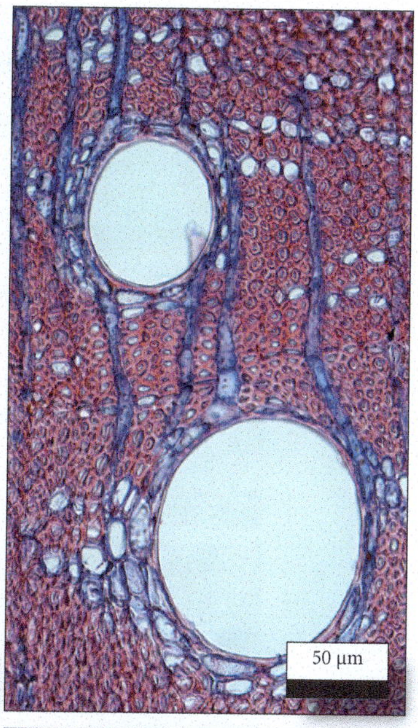

STEM XYLEM ANATOMY

Transverse section

Growth ring boundaries indistinct or absent.
Wood diffuse-porous.
Vessels solitary or in radial multiples of 2 to 5 elements.
Tangential diameter of vessels 110-134-180 µm, 16-20-33 vessels/mm^2.
Gums and/or other deposits in heartwood vessels.
Fibres very thick-walled.
Apotracheal parenchyma diffuse-in-aggregates.
Paratracheal parenchyma vasicentric to aliform.
Rays 20-21-22/mm.

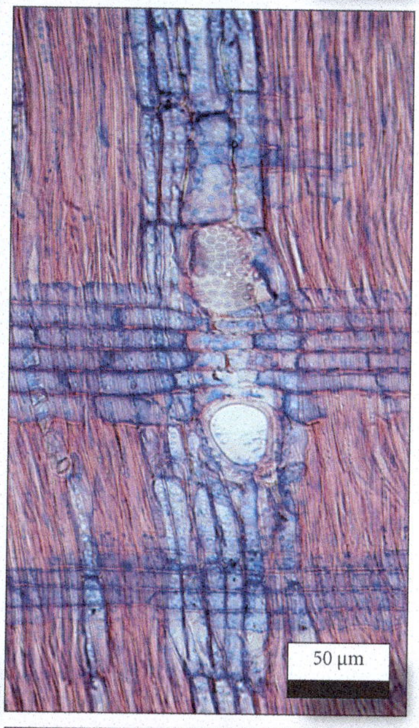

Radial section

Simple perforation plates.
Intervessel pits oval to circular and alternate, 4-7 µm in diameter.
Vestured pits present.
Vessel-ray pits with distinct borders; similar to intervessel pits in size and shape throughout the ray cell.
Vessel element length 150-167-190 µm.
Fibres with simple to minutely bordered pits.
Homogeneous and heterogeneous rays both present. A few rays with body ray cells procumbent and one row of square and upright cells.
Prismatic crystal in chambered axial parenchyma cells.

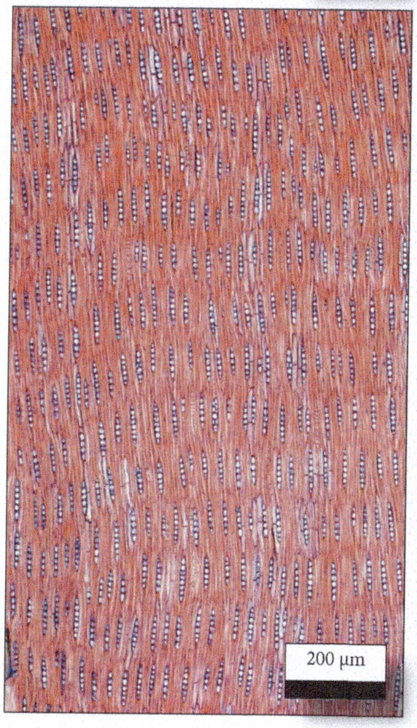

Tangential section

Rays exclusively uniseriate.
Ray height 80-117-160 µm.
Fusiform parenchyma cells present. Two cells per parenchyma strand.
Rays, vessels and axial parenchyma regularly storied. Number of ray tiers per axial mm: 15-16.

Dalbergia purpurascens Baill.

EN: Rosewood FR: Bois de rose ML: Voamboana

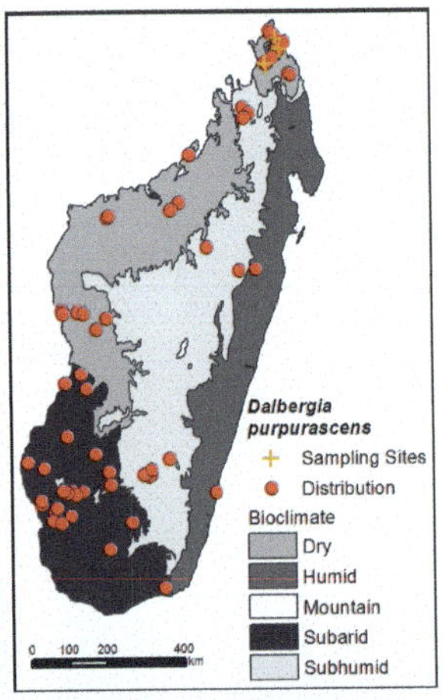

PLANT DESCRIPTION

Tree | 8-25 m high | 20-80 cm DBH
Endemic to Madagascar. Exploited species.

DISTRIBUTION

Elevation: 0-1000 m a.s.l.
Vegetation formation: Forest and woodland
Bioclimate: Dry, sub-arid and sub-humid
Protected areas: Andringitra, Ankarafantsika, Isalo, Ankarana, Bemaraha, Kirindy-Mitea, Namoroka, Zahamena and Zombitsy-Vohibasia
New protected areas: Montagne des Français and Sahafary

Slides made from material collected in Ankarana.

PITH AND PRIMARY XYLEM ANATOMY

Pith: Shape of pith round. Cells dimorphic. Prismatic crystals present. Simple pits in transverse cell walls. Pith not visible in polarized light.

Primary xylem: Vascular bundles not distinct. Vessels solitary and in short radial multiples. Fibres thin- to thick-walled in radial rows.

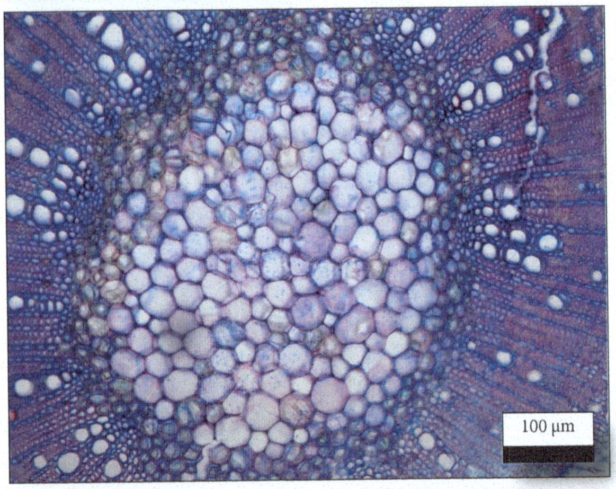

BRANCH XYLEM ANATOMY

Description: Growth ring boundaries distinct. Wood ring-porous. Tangential diameter of earlywood vessels 50-80-120 µm and latewood vessels 20-39-50 µm. Vessel element length 70-106-130 µm. Intervessel pits small, 4-7 µm in diameter. Fibres thick- to very thick-walled. Axial parenchyma diffuse-in-aggregates and in narrow lines up to 3 cells. Rays mostly biseriate. Ray height 80-106-160 µm, rays 10-13-18/mm.

Anatomical differences to stem xylem: Earlywood and latewood vessels are smaller in branch (50-80-120/130-219-360 µm; 20-39-50/50-77-120 µm). Rays are shorter in branch (80-106-160/90-166-250 µm) while ray number is higher in branch (10-13-18/7-9-11 rays/mm).

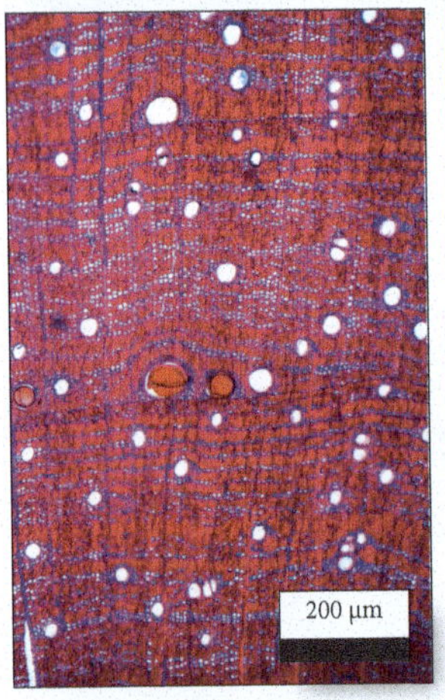

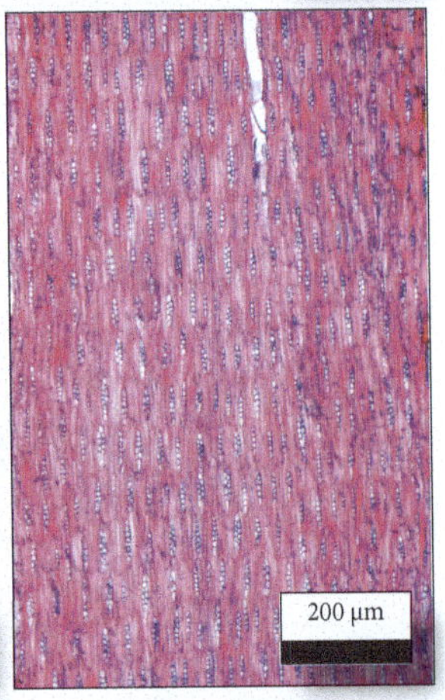

BARK ANATOMY

Branch bark: Collapsed sieve tubes and groups of fibres tangentially distributed in the phloem. Prismatic crystals present. The cortex consists of irregularly blue-stained parenchyma cells with or without prismatic crystals.

Stem bark: The phloem consists of alternating bands of fibres and collapsed and non-collapsed sieve tubes. Groups of fibres irregularly dispersed in the phloem. Parenchyma dilatation and prismatic crystals present. The phellem is homogeneous, it consists of regularly arranged rectangular cells. Lignified cells in phellem, distinct in polarized light.

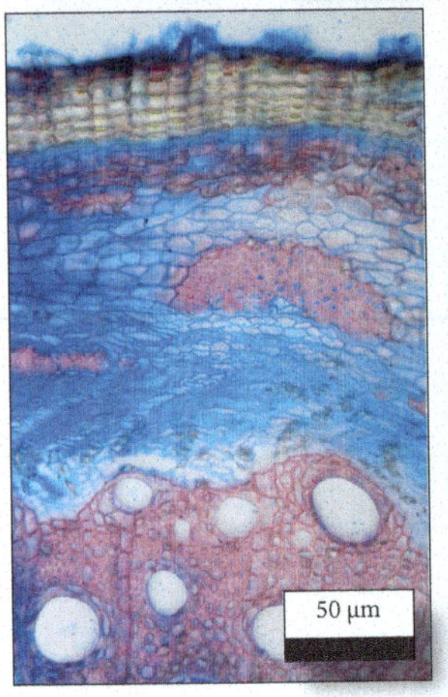

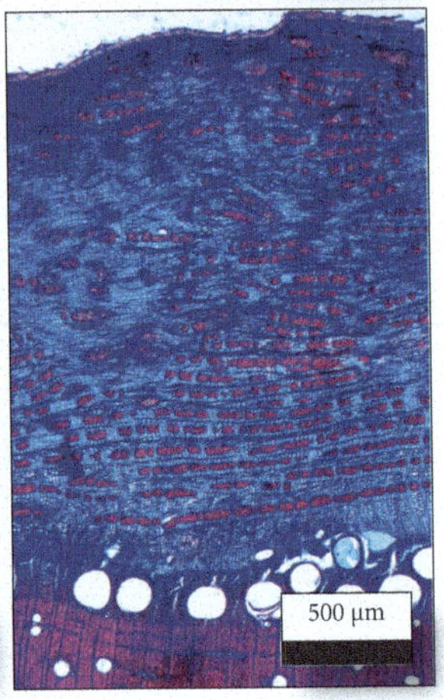

IAWA feature numbers

1 3 9 13 22 26 29 30 42 43 47 52 58 61 66 70 77 89 90 91 97 102.2 106 115 118 120 136 142 180 189 196

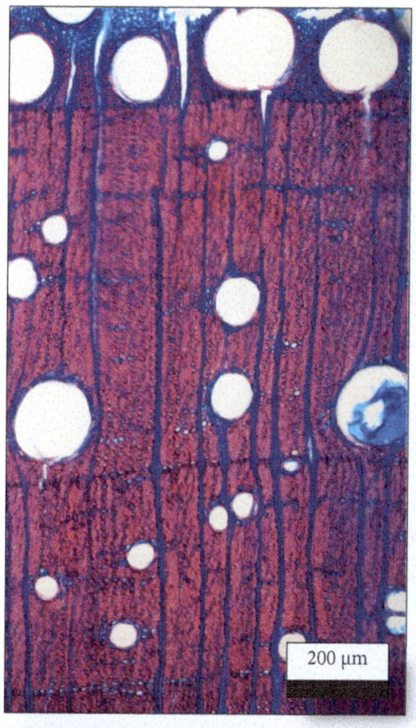

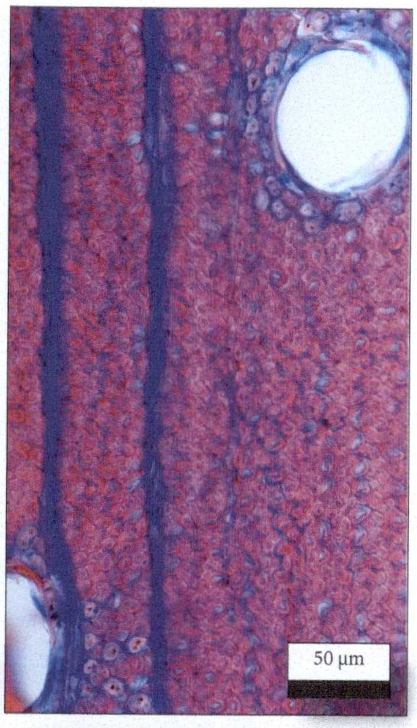

STEM XYLEM ANATOMY

Transverse section

Growth ring boundaries distinct, marked by thick-walled and radially flattened latewood fibres.
Wood ring-porous.
Vessels mostly solitary.
Tangential diameter of earlywood vessels 130-219-360 µm and latewood vessels 50-77-120 µm, 4-7-16 latewood vessels/mm^2.
Gums and/or other deposits in heartwood vessels.
Fibres very thick-walled.
Apotracheal parenchyma diffuse and diffuse-in-aggregates. Marginal or seemingly marginal parenchyma bands.
Rays 7-9-11/mm.

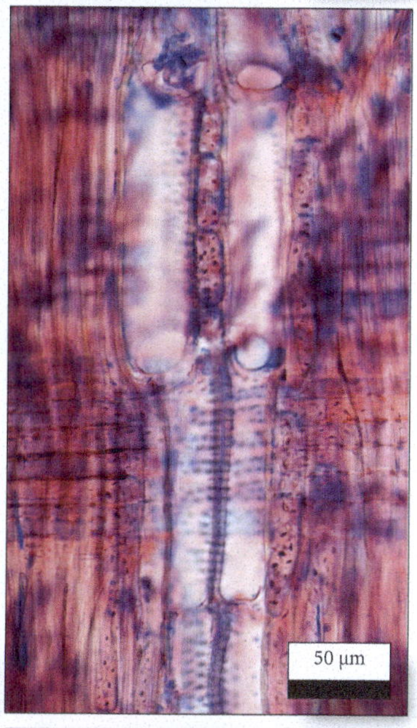

Radial section

Simple perforation plates.
Intervessel pits oval to circular and alternate, 4-7 µm in diameter.
Vestured pits present.
Vessel-ray pits with distinct borders; similar to intervessel pits in size and shape throughout the ray cell.
Vessel element length 170-191-240 µm.
Fibres with simple to minutely bordered pits.
Heterogeneous rays with body ray cells procumbent and one row of upright and square marginal cells.
Prismatic crystal in chambered axial parenchyma cells.

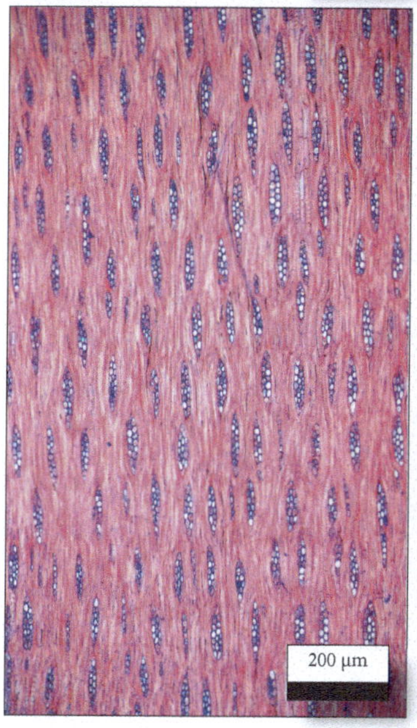

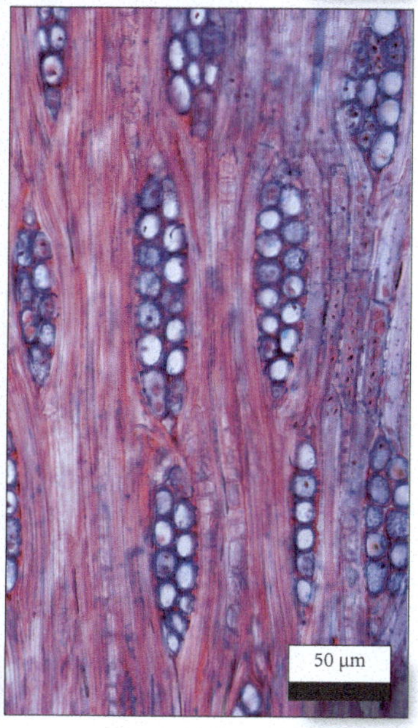

Tangential section

Rays mostly biseriate.
Ray height 90-166-250 µm.
Fusiform parenchyma cells present. Two cells per parenchyma strand. Rays, vessels and axial parenchyma storied. Number of ray tiers per axial mm: 13-14.

Dalbergia suaresensis Baill.

Synonymy: *D. bernieri* Baill.

EN: Palisandre FR: Palissandre ML: Manary

PLANT DESCRIPTION

Tree | 6-8 m high | 8-40 cm DBH
Endemic to Madagascar. Exploited species.

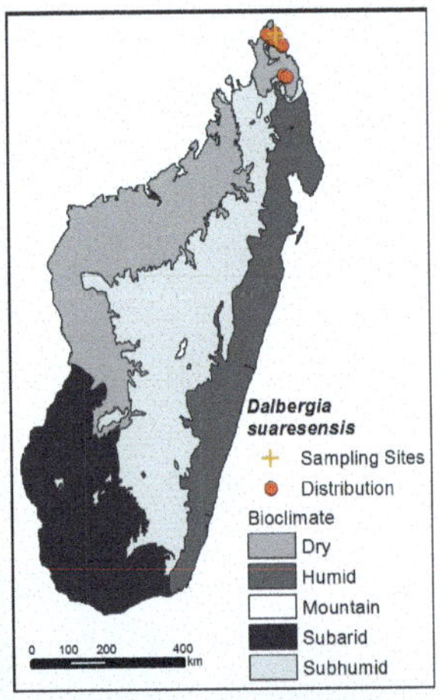

DISTRIBUTION

Elevation: 0-500 m a.s.l.
Vegetation formation: Forest
Bioclimate: Dry
New protected areas: Montagne des Français and Orangea/ Rigny

Slides made from material collected in Montagne des Français.

PITH AND PRIMARY XYLEM ANATOMY

Pith: Shape of pith round. Thick-walled parenchyma cells present. Cells dimorphic. Dark staining substances rare. Prismatic crystals present. Simple pits in transverse cell walls. Pith not visible in polarized light.

Primary xylem: Vascular bundles not distinct. Vessels solitary and in short radial multiples.

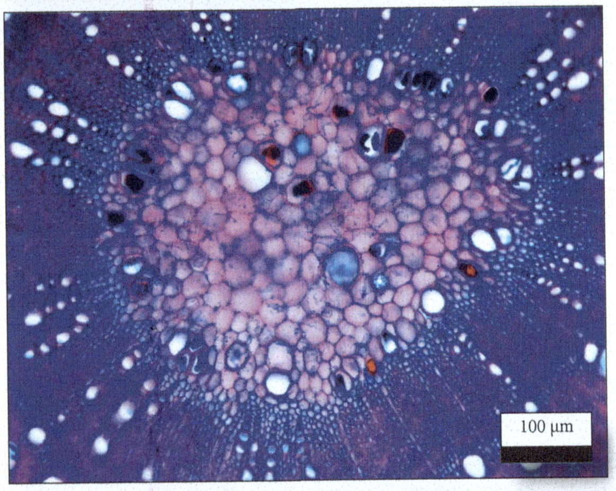

BRANCH XYLEM ANATOMY

Description: Growth rings boundaries indistinct or absent. Vessels of two distinct diameter classes. Vessels diameter 30-75-140 μm, 21-34-75 vessels/mm^2. Vessel elements length 100-135-160 μm. Intervessel pits small, 4-7 μm in diameter. Fibres thick- to very thick-walled. Axial parenchyma in narrow lines up to 3 cells. Rays predominantly uniseriate. Ray height 40-108-250 μm, rays 8-10-12/mm.

Anatomical differences to stem xylem: Vessels diameter is smaller in branch (30-75-140/60-120-240 μm). Vessel density is higher in branch (21-34-75/5-9-15 vessels/mm^2). Ray height and number are slightly lower in the branch (40-108-250/60-160-360 μm; 8-10-12/10-12-14 rays/mm).

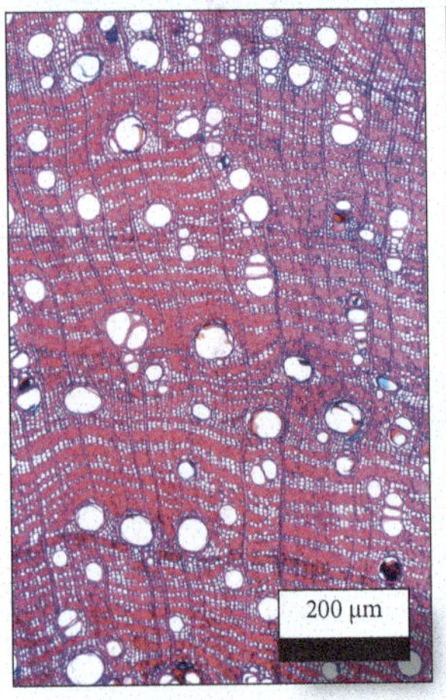

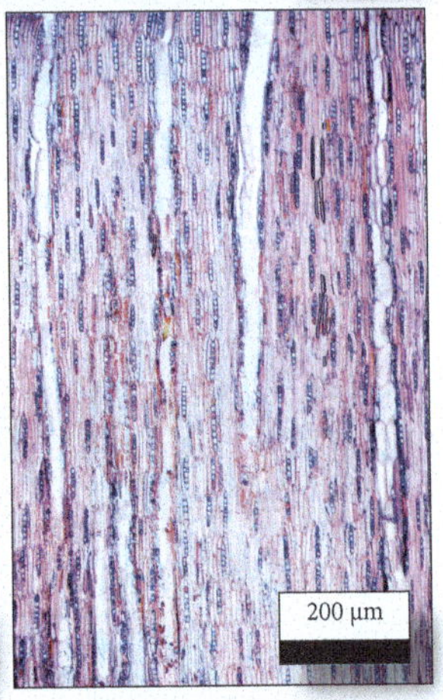

BARK ANATOMY

Branch bark: Collapsed sieve tubes and prismatic crystals present in the phloem. The cortex consists on irregularly blue-stained parenchyma cells and groups of tangentially distributed fibres.

Stem bark: Phloem consists of alternating fibre bands and collapsed and non-collapsed sieve tubes. The cortex is characterized by groups of sclereids tangentially distributed. Axial parenchyma dilatation present. Prismatic crystals present in the sclereids. The phellem is homogeneous, it consists of regularly arranged rectangular cells. Lignified cells in phellem, distinct in polarized light.

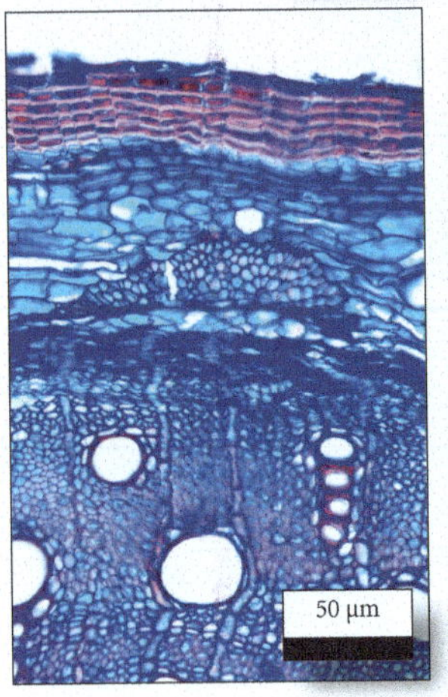

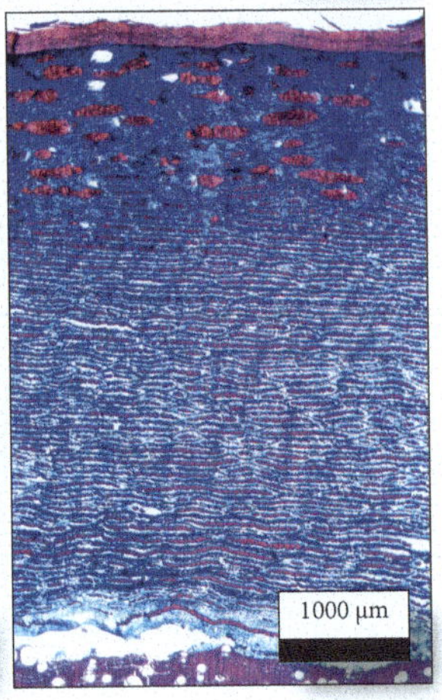

IAWA feature numbers
2 5 9 13 22 25 29 30 42 45 47 52 58 61 66 70 77 79 86 90
91 96.1 102.2 104 116 118 120 136 142 180 189 192 196

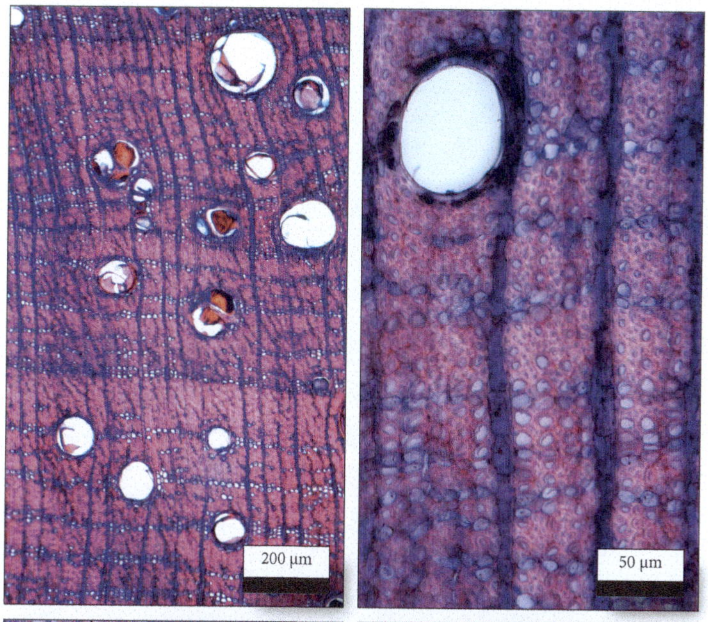

STEM XYLEM ANATOMY

Transverse section

Growth rings boundaries indistinct or absent.
Wood diffuse-porous.
Vessels solitary or in radial multiples of 2-6 elements.
Vessels of two distinct diameter classes.
Tangential diameter of vessels 60-120-240 µm, 5-9-15 vessels/mm^2.
Gums and/or other deposits in heartwood vessels.
Fibres very thick-walled.
Apotracheal parenchyma diffuse-in-aggregates, and in narrow lines up to 3 cells. Paratracheal parenchyma vasicentric.
Rays 10-12-14/mm.

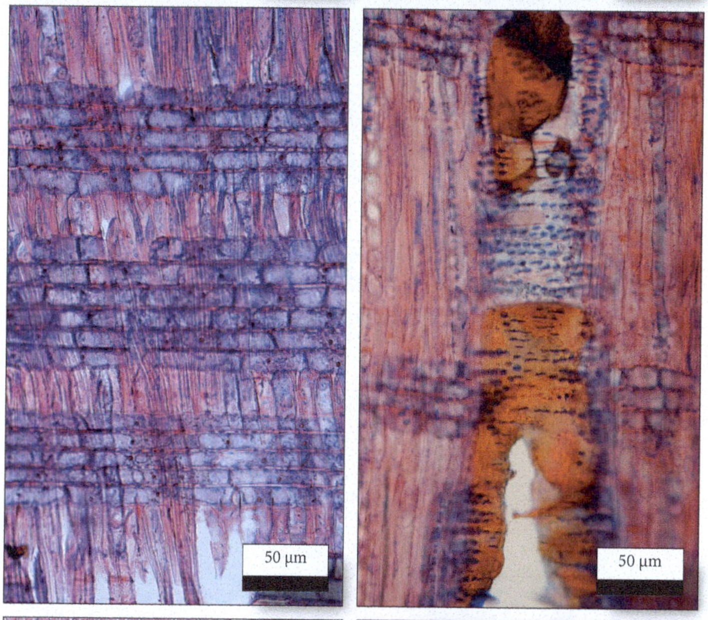

Radial section

Simple perforation plates.
Intervessel pits oval to circular and opposite, 4-7 µm in diameter.
Vestured pits present.
Vessel-ray pits with distinct borders; similar to intervessel pits in size and shape throughout the ray cell.
Vessel element length 90-156-300 µm.
Fibres with simple to minutely bordered pits.
Homogeneous rays with all ray cells procumbent.
Prismatic crystal in chambered axial parenchyma cells.

Tangential section

Rays predominantly uniseriate. Ray height 60-110-360 µm.
Fusiform parenchyma cells present. Two cells per parenchyma strand.
Rays, vessels and axial parenchyma irregularly storied. Number of ray tiers per axial mm: 16-17.

Dalbergia trichocarpa Baker

Synonymy: *D. boinensis* Jum., *D. perrieri* Drake

EN: Palisander FR: Palissandre ML: Manary

PLANT DESCRIPTION

Tree | 4-25 m high | 15-40 cm DBH
Endemic to Madagascar. Exploited species.

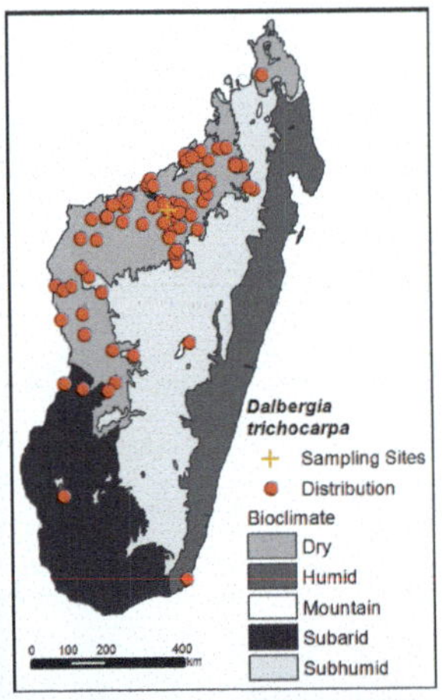

DISTRIBUTION

Elevation: 0-1000 m a.s.l.
Vegetation formation: Forest, grassland and woodland
Bioclimate: Dry and sub-arid
Protected areas: Ankarafantsika, Bemaraha and Namoroka

Slides made from material collected in Ankarafantsika.

PITH AND PRIMARY XYLEM ANATOMY

Pith: Shape of pith round. Layer of blue-stained cells surround the pith. Thick-walled parenchyma cells present. Cells dimorphic. Dark staining substances abundant. Simple pits in transverse cell walls. Pith not visible in polarized light.

Primary xylem: Vascular bundles not distinct. Vessels solitary and in short radial multiples.

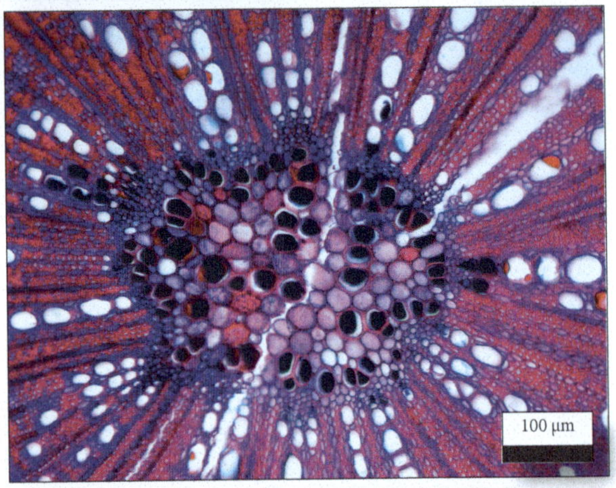

BRANCH XYLEM ANATOMY

Description: Growth rings boundaries indistinct or absent. Tangential diameter of vessels 60-103-150 µm, 24-31-50 vessels/mm^2. Vessel elements length 130-144-160 µm. Intervessel pits small, 4-7 µm in diameter. Fibres thin- to thick-walled. Axial parenchyma in bands 2 to 5 cells wide. Rays predominantly uniseriate. Ray height 60-96-120 µm. Rays 10-14-18/mm.

Anatomical differences to stem xylem: Mean tangential diameter of vessel lumina is slightly smaller in branch (60-103-150/120-174-260 µm). Vessel density is higher in the branch (24-31-50/5-9-22 vessels/mm^2). Rays are shorter in the branch (60-96-120/130-144-160 µm) while ray number is similar (10-14-18/11-14-17 rays/mm).

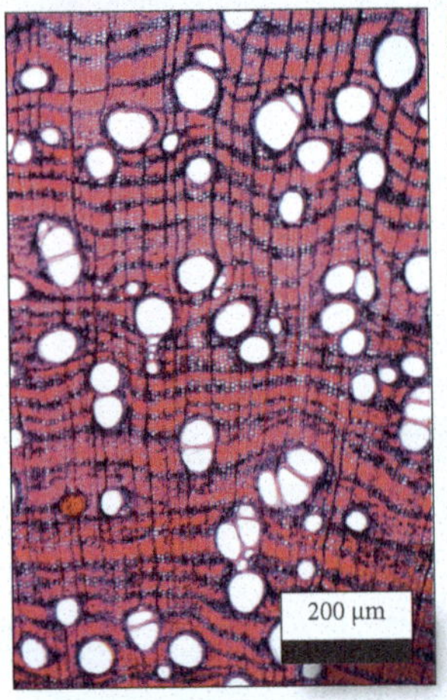

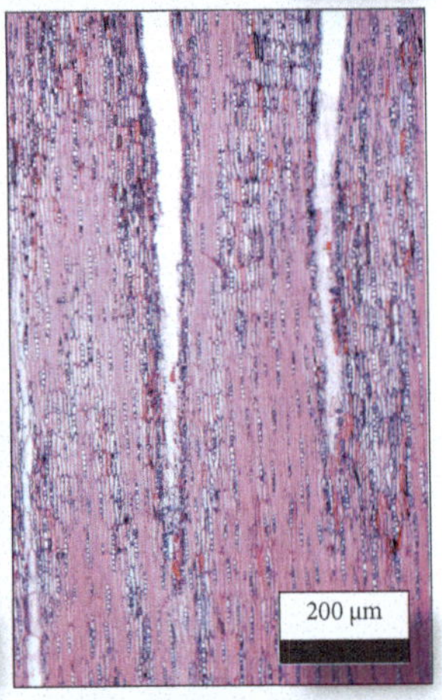

BARK ANATOMY

Branch bark: The phloem consists of alternating bands of fibres and collapsed and non-collapsed sieve tubes. Prismatic crystals present. The cortex consists of irregular rectangular blue-stained parenchyma cells and tangential groups of fibres.

Stem bark: The phloem consists of alternating tangential bands of fibres and collapsed and non-collapsed sieve tubes. Parenchyma dilatation present and crystal sand scattered or irregularly dispersed in the phloem. The phellem is homogeneous, consists of regularly arranged rectangular parenchyma cells with rarely dark staining substances. Lignified cells in phellem, distinct in polarized light.

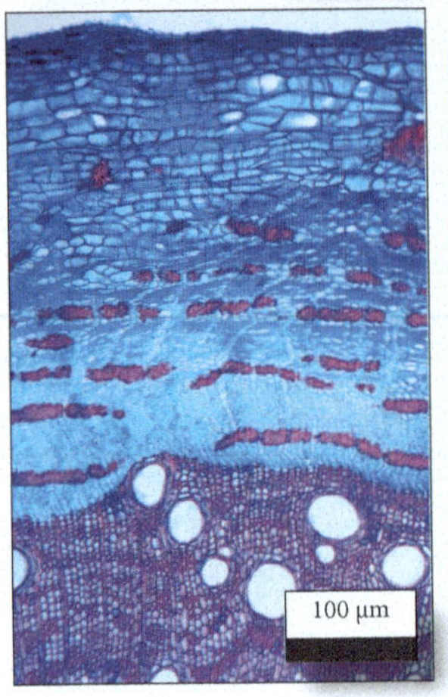

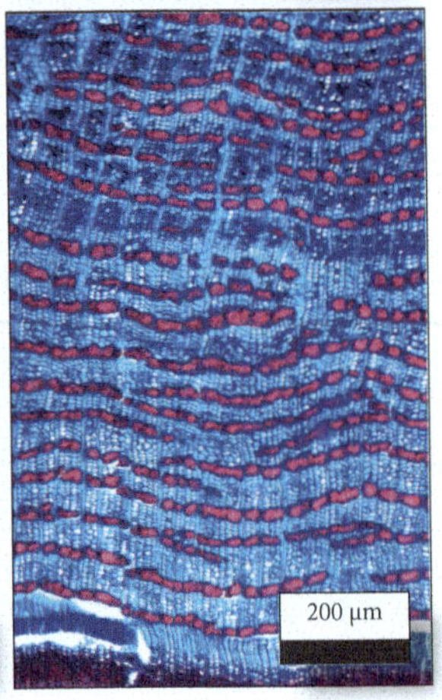

IAWA feature numbers
2 5 9 13 22 25 29 30 42 47 52 58 61 66 70 78 79 86 90 91 96.1 102.1 104 116 118 120 136 142 180 189 192 196

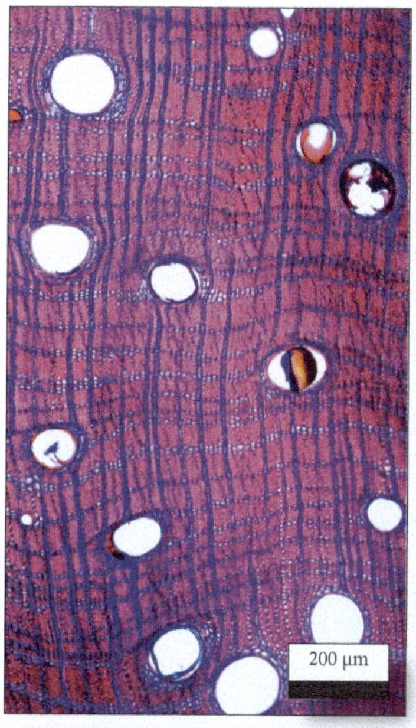

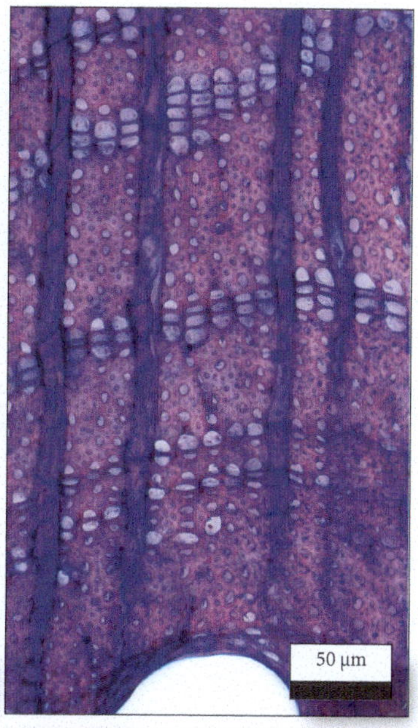

STEM XYLEM ANATOMY

Transverse section

Growth rings boundaries indistinct or absent.
Wood diffuse-porous.
Vessels mostly solitary.
Tangential diameter of vessels 120-174-260 µm, 5-9-22 vessels/mm².
Gums and/or other deposits in heartwood vessels.
Fibres very thick-walled.
Apotracheal parenchyma in narrows lines up to 3 cells wide. Paratracheal parenchyma vasicentric.
Rays 11-14-17/mm.

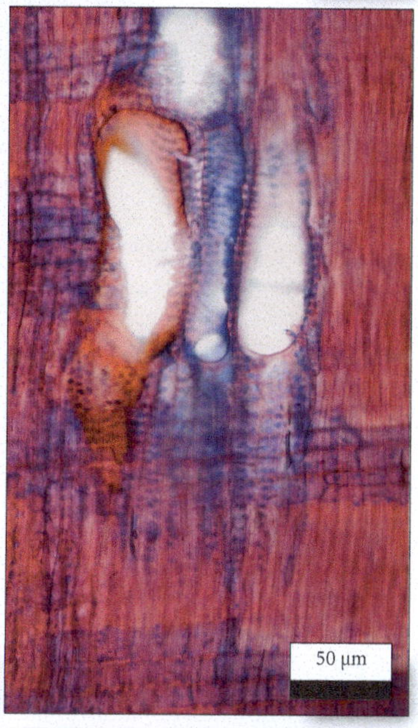

Radial section

Simple perforation plates.
Intervessel pits oval to circular and alternate, 4-7 µm in diameter.
Vestured pits.
Vessel-ray pits with distinct borders; similar to intervessel pits in size and shape throughout the ray cell.
Vessel element length 130-144-160 µm.
Fibres with simple to minutely bordered pits (libriform fibres).
Homogeneous rays, all ray cells procumbent.
Prismatic crystal in chambered axial parenchyma cells.

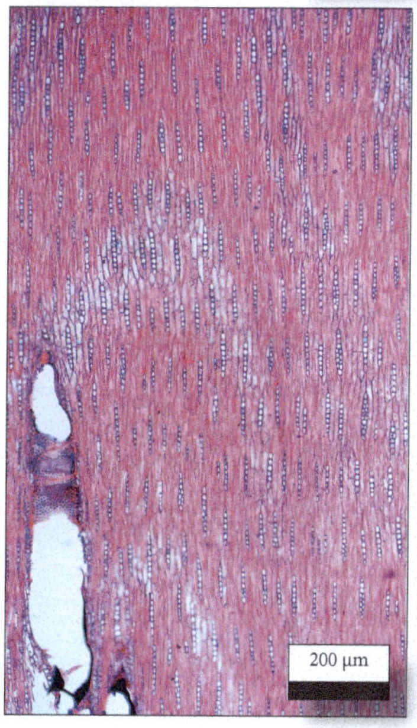

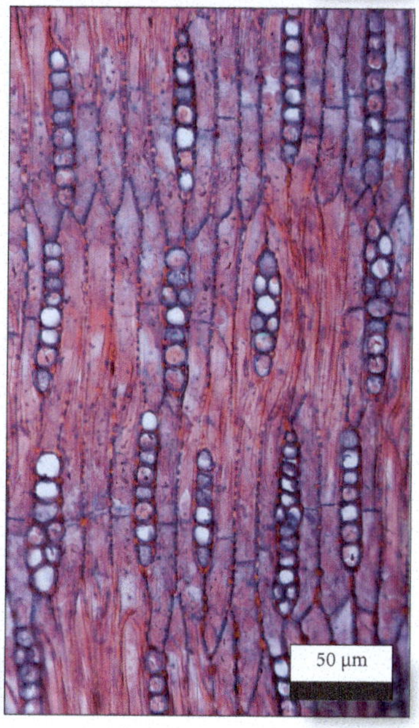

Tangential section

Rays uni- and biseriate, predominantly uniseriate.
Ray height 60-128-240 µm.
Fusiform parenchyma cells present. Two cells per parenchyma strand.
Rays and axial parenchyma storied. Number of ray tiers per axial mm: 15-16.

6. *Diospyros*

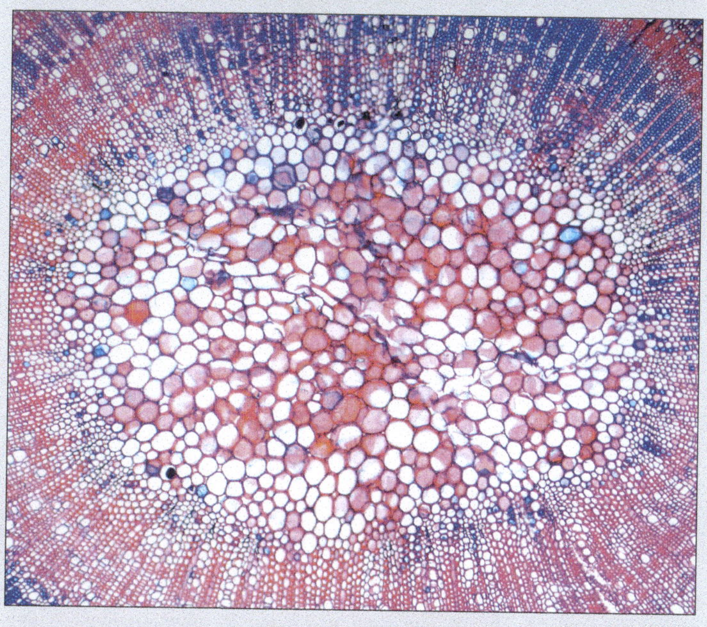

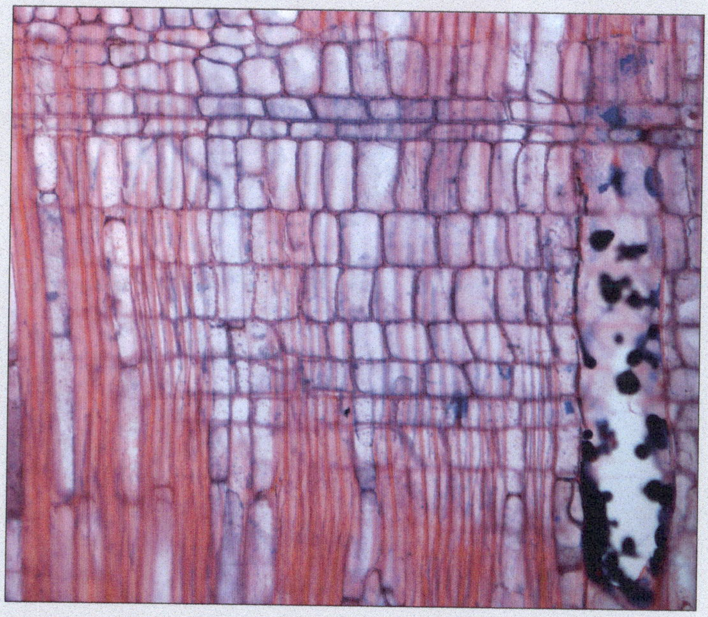

The taxonomy of *Diospyros* is currently being revised. A total of 85 described species are currently recognized, and at least 130 additional species remain to be named and described.
Here we describe the stem anatomy of 31 *Diospyros* species.

B.H. Ravaomanalina et al., *Stem Anatomy of Dalbergia and Diospyros Species from Madagascar*, DOI 10.1007/978-3-319-51147-4_6

Diospyros aculeata H. Perrier.

Synonymy: *D. aculeata* subsp. *aculeata* H. Perrier, *D. aculeata* subsp. *meridionalis* H. Perrier

EN: Ebony FR: Bois d'ébène ML: Hazomainty

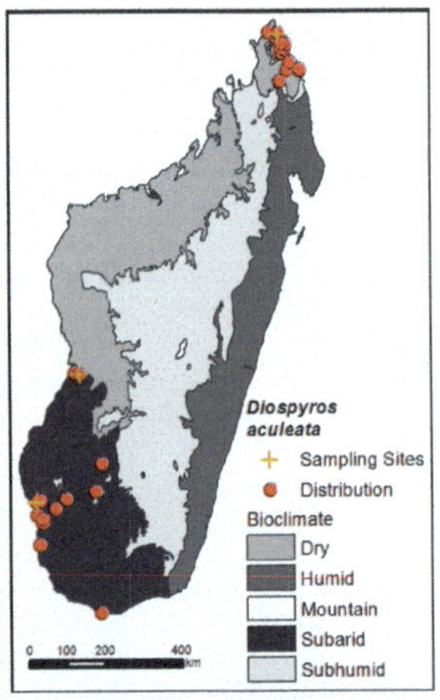

PLANT DESCRIPTION

Shrub to tree | 3-8 m | 20-30 cm DBH
Endemic to Madagascar. Exploited species.

DISTRIBUTION

Elevation: 0-500 m a.s.l.
Vegetation formation: Forest
Bioclimate: Dry and sub-arid
Protected areas: Analamerana, Kirindy Menabe, Cap Ste Marie and Isalo
New protected areas: Daraina complex, La Table/St Augustin, Montagne des Français, Orangea/Rigny, Sahafary and Ranobe
Slides made from material collected in Montagne des Français.

PITH AND PRIMARY XYLEM ANATOMY

Pith: Shape of pith pentagonal. Thick-walled parenchyma cells present. Cells dimorphic. Prismatic crystals present. Pith visible in polarized light.

Primary xylem: Vascular bundles not distinct. Vessels solitary and in radial multiples.

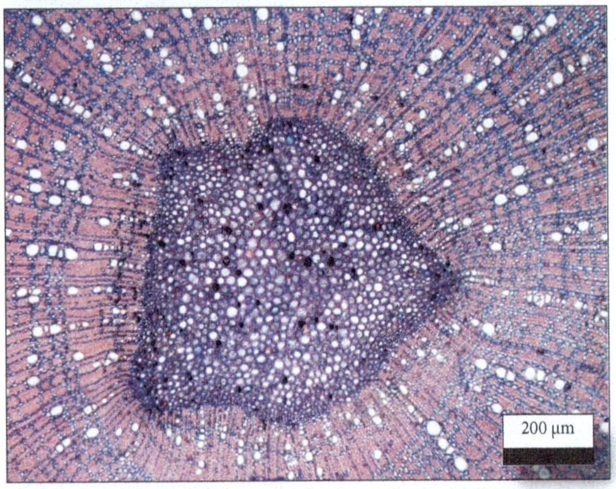

BRANCH XYLEM ANATOMY

Description: Growth rings boundaries indistinct or absent. Tangential diameter of vessels 20-35-50 µm, 60-73-82 vessels/mm^2. Vessel element length 120-250-420 µm. Intervessel pits small, 4-7 µm in diameter. Fibres thin- to thick-walled. Axial parenchyma diffuse-in-aggregates. Rays exclusively uniseriate, ray height 170-365-760 µm, 38-42-46 rays/mm.

Anatomical differences to stem xylem: Mean tangential diameter of vessel lumina is smaller in the branch (20-35-50/40-51-60 µm), vessel density is higher in the branch (60-73-82/22-39-51 vessels/mm^2). Ray height is lower in the branch (170-365-760/100-397-780 µm), why rays are more numerous in branch (38-42-46/22-25-29 rays/mm).

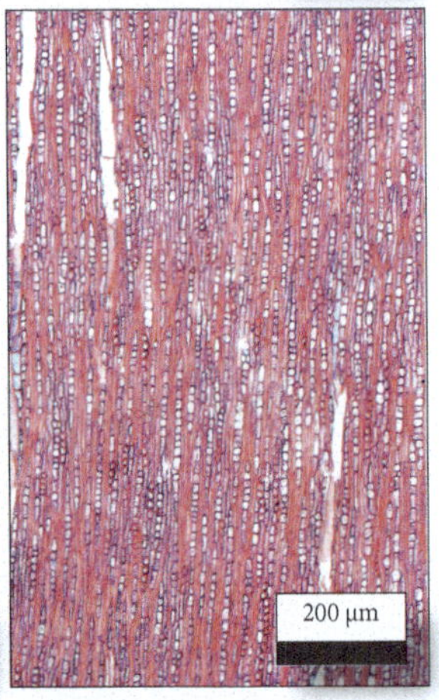

BARK ANATOMY

Branch bark: Collapsed and non-collapsed sieve tubes present in the phloem. Groups of dispersed sclereids present. Some rays become dilated. The cortex is characterized by a band of sclereids.

Stem bark: The phloem is composed only of sieve tubes and parenchyma cells. The cortex is defined by a band of sclereids. Layered phellem. The phellem cells are homogeneous and it consist of regularly arranged rectangular cells, indistinct in polarized light.

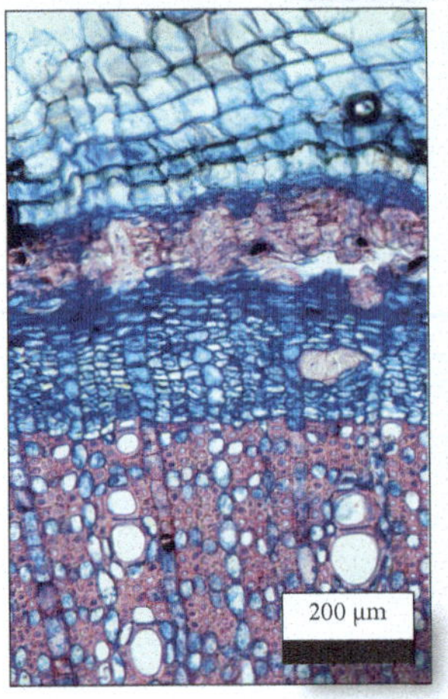

IAWA feature numbers

2 5 9 9.1 13 22 25 30 41 48 52 58 61 66 70 77 78 92
93 96 102.3 108 116 136 138 142 180 189 190 192 196

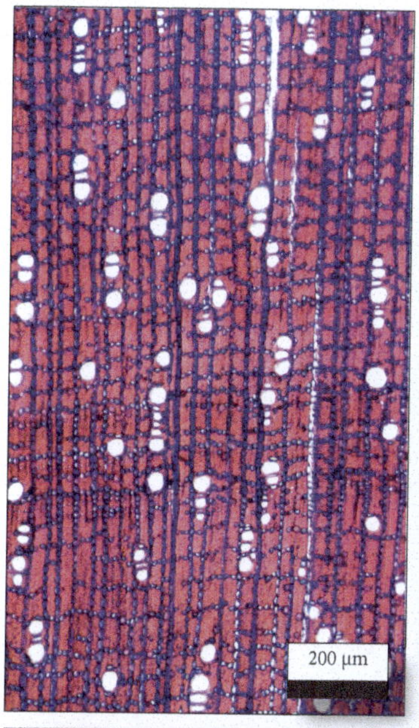

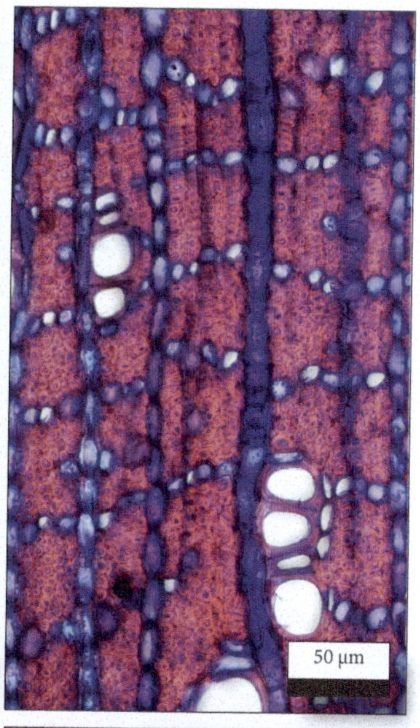

STEM XYLEM ANATOMY

Transverse section

Growth rings boundaries indistinct or absent.
Wood diffuse-porous.
Vessels solitary and in short radial multiples of 2 to 4 elements.
Tangential diameter of vessels medium 40-51-60 µm, 22-39-51 vessels/mm^2.
Gums and/or other deposits in heartwood vessels.
Fibers very thick-walled.
Apotracheal parenchyma diffuse-in-aggregates.
Paratracheal parenchyma scanty.
Rays 22-25-29/mm.

Radial section

Simple perforation plates.
Intervessel pits circular and alternate, 4-7 µm in diameter.
Vessel-ray pits with distinct borders; similar to intervessel pits in size and shape throughout the ray cell.
Vessel element length 180-282-420 µm.
Fibres with simple to minutely bordered pits (libriform fibers).
Heterogeneous rays, body ray cells procumbent, with 2 to over 4 rows of upright and square marginal cells.
Prismatic crystals in procumbent ray cells.

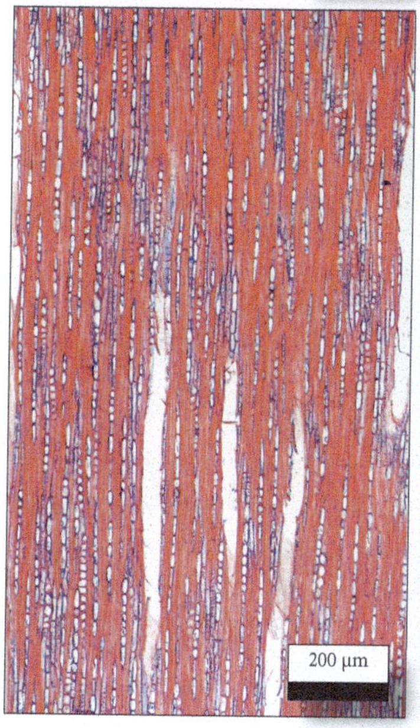

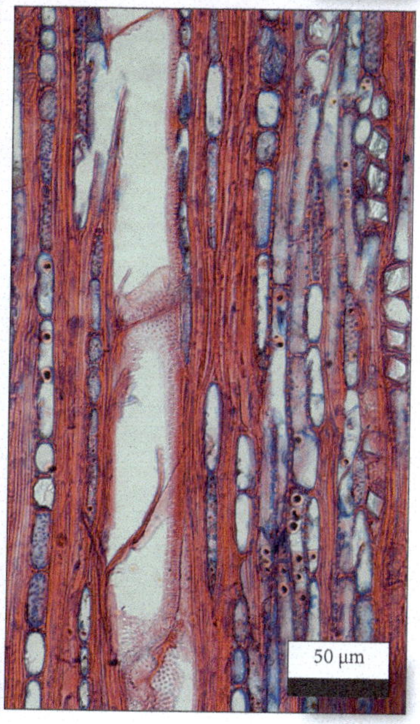

Tangential section

Rays exclusively uniseriate.
Ray height 100-397-780 µm.
Three to 8 cells per parenchyma strand.

Diospyros acutiflora ined.

EN: Ebony FR: Bois d'ébène ML: Hazomainty

PLANT DESCRIPTION

Shrub to tree | 7-25 m high | 17 cm DBH
Endemic to Madagascar. Exploited species.

DISTRIBUTION

Elevation: 0-500 m a.s.l.
Vegetation formation: Forest
Bioclimate: Humid
Protected areas: Betampona
New protected areas: Mahabo-Mananivo, Makirovana, Analalava and Tampolo

Slides made from material collected in Analalava.

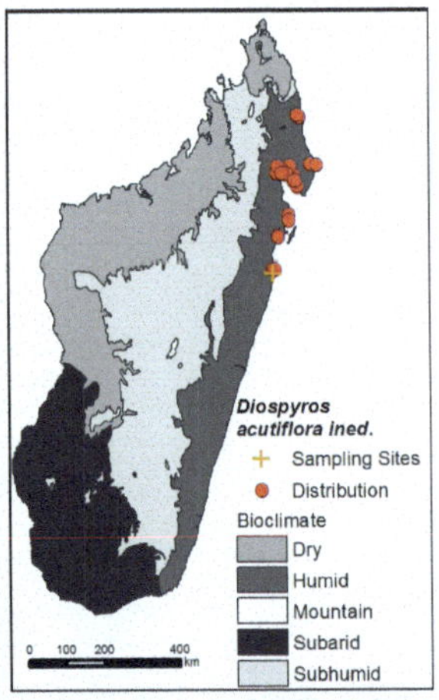

PITH AND PRIMARY XYLEM ANATOMY

Pith: Shape of pith round. Cells dimorphic. Pits in transverse cell walls. Pith not visible in polarized light.

Primary xylem: Vascular bundles separated. Vessels solitary and in short radial multiples.

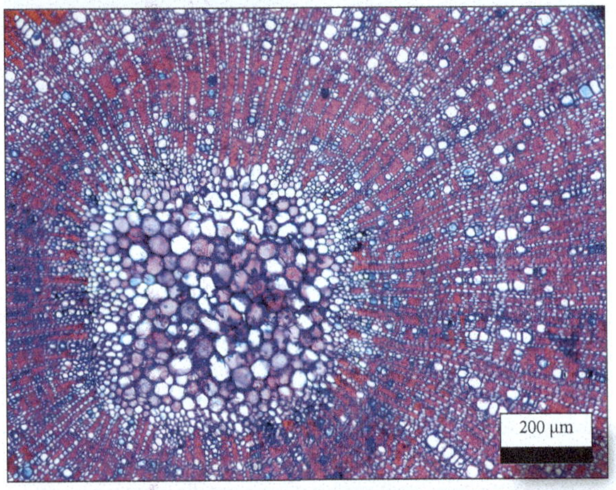

BRANCH XYLEM ANATOMY

Description: Growth rings boundaries indistinct or absent. Tangential diameter of vessels 30-43-60 µm, 39-56-81 vessels/mm^2. Vessel element length 200-376-490 µm. Intervessel pits small, 4-7 µm in diameter. Fibres thick- to very thick-walled. Axial parenchyma diffuse-in-aggregates. Rays exclusively uniseriate. Ray height 450-741-1350 µm. Rays 35-38-44/mm.

Anatomical differences to stem xylem: Mean tangential diameter of vessel lumina is smaller in the branch (30-43-60/70-78-100 µm), while vessel density is higher (39-56-81/17-34-45 vessels/mm^2). Ray number is slightly higher in the branch (35-38-44/27-31-34 rays/mm).

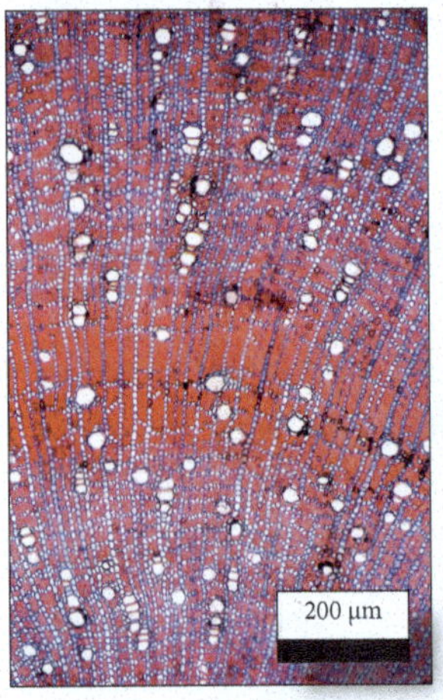

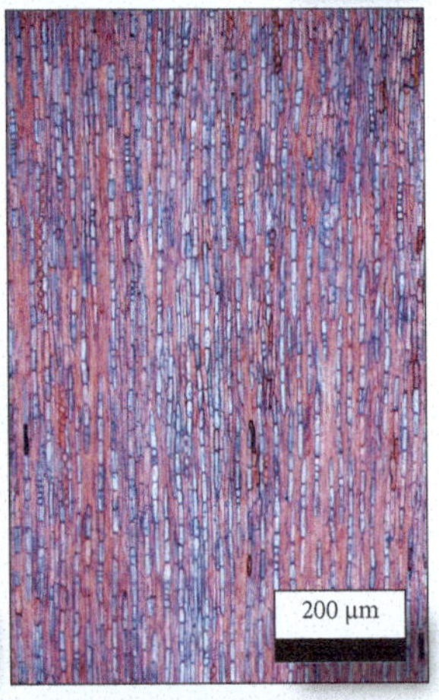

BARK ANATOMY

Branch bark: Collapsed and non-collapsed sieve tubes present in the phloem. The cortex is presented by a band of sclereids and irregularly blue-stained parenchyma cells.

Stem bark: The phloem consists of collapsed and non-collapsed sieve tubes, and irregularly distributed groups of sclereids. Ray dilatation present. The cortex is characterized by bands of sclereids. The phellem is homogeneous with dark stained substances and it consists of regularly arranged rectangular cells, indistinct in polarized light.

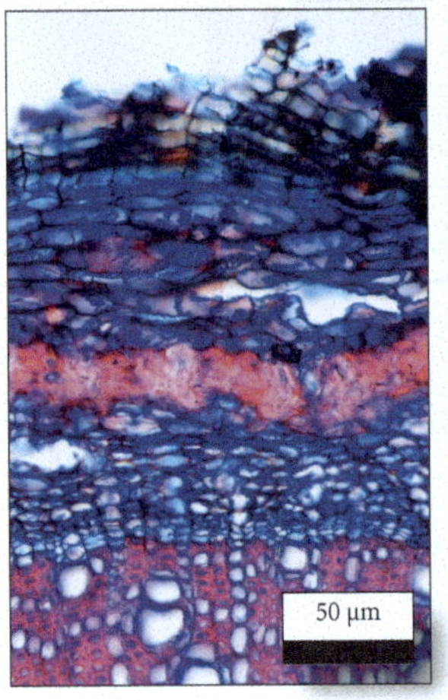

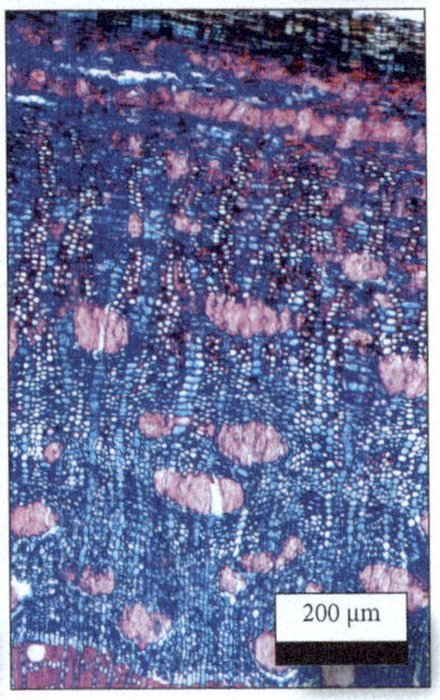

IAWA feature numbers

2 5 9 9.1 13 22 25 30 41 48 53 58 61 66 70 77 78 92 96
102.4 107 108 116 136 138 142 180 189 192 196

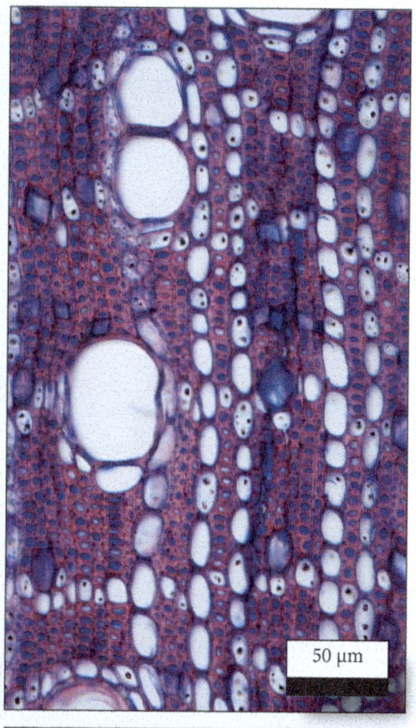

STEM XYLEM ANATOMY

Transverse section

Growth rings boundaries indistinct or absent.
Wood diffuse-porous.
Vessels solitary and in radial multiples of 2 to 5 elements.
Tangential diameter of vessels medium 70-78-100 µm, 17-34-45 vessels/mm^2.
Gums and/or other deposits in heartwood vessels.
Fibers very thick-walled.
Apotracheal parenchyma diffuse-in-aggregates.
Paratracheal parenchyma scanty.
Rays 27-31-34/mm.

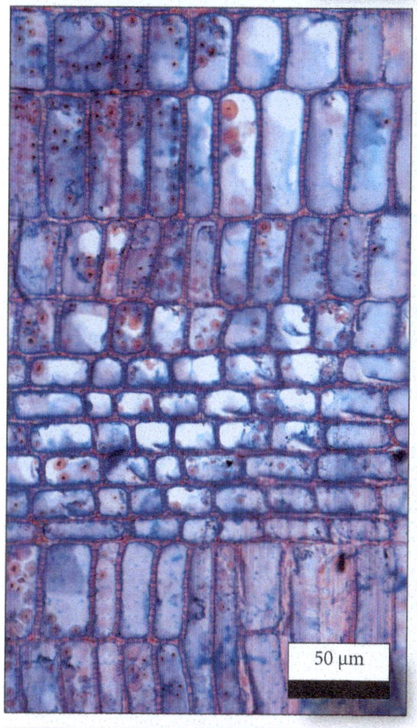

Radial section

Simple perforation plates.
Intervessel pits circular and alternate, 4-7 µm in diameter.
Vessel-ray pits with distinct borders; similar to intervessel pits in size and shape throughout the ray cell.
Vessel element length 200-427-620 µm.
Fibres with simple to minutely bordered pits (libriform fibers).
Heterogeneous rays, body ray cells procumbent with 2 to over 4 rows of upright and square marginal cells.
Prismatic crystal in chambered axial parenchyma cells. Few prismatic crystals in procumbent ray cells.

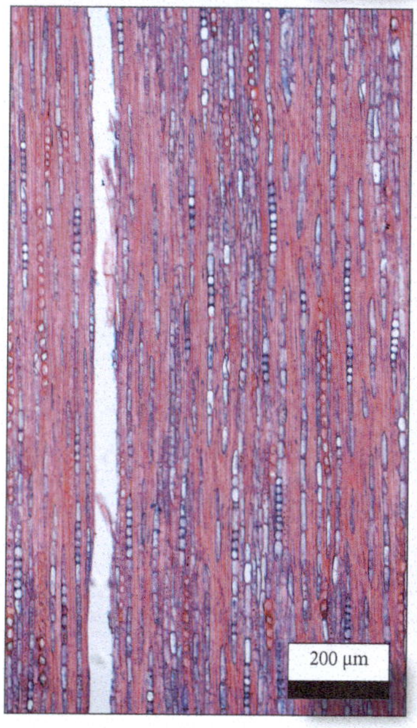

Tangential section

Rays exclusively uniseriate.
Ray height 410-697-1400 µm.
Two to 4 cells per parenchyma strand.

Diospyros analamerensis H. Perrier.

EN: Ebony   FR: Bois d'ébène   ML: Hazomainty

PLANT DESCRIPTION

Tree | 4-15 m high | 16-18 cm DBH
Endemic to Madagascar. Non-exploited species.

DISTRIBUTION

Elevation: 0-500 m a.s.l.
Vegetation formation: Forest
Bioclimate: Dry
Protected areas: Analamerana and Ankarana
New protected areas: Daraina complex, Montagne des Français and Sahafary

Slides made from material collected in Montagne des Français.

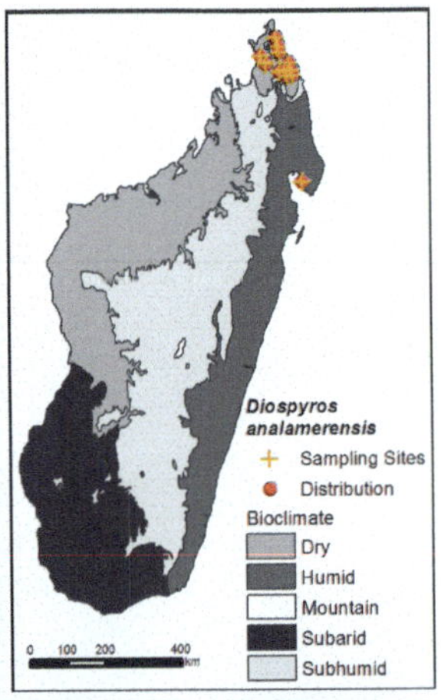

PITH AND PRIMARY XYLEM ANATOMY

Pith: Shape of pith oval. Blue-stained cells present. Cells dimorphic. Dark staining substances present. Prismatic crystals present. Pith visible in polarized light.

Primary xylem: Vascular bundles not distinct. Vessels solitary and in short radial multiples.

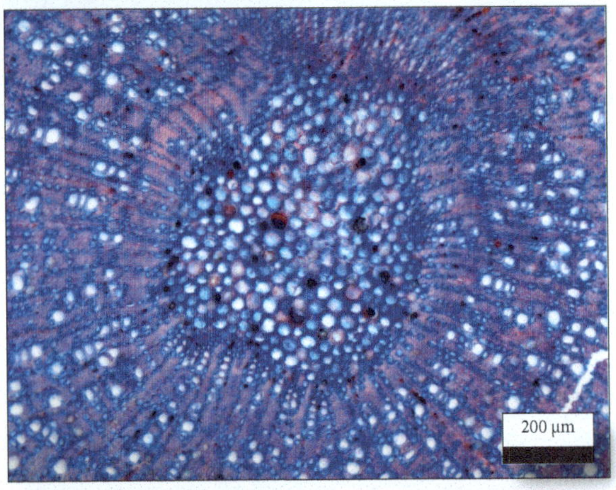

BRANCH XYLEM ANATOMY

Description: Growth rings boundaries indistinct or absent. Tangential diameter of vessels 10-29-40 µm, 105-125-140 vessels/mm². Vessel element length 120-298-400 µm. Intervessel pits small, 4-7 µm in diameter. Fibres thin- to thick-walled. Axial parenchyma diffuse-in-aggregates. Rays exclusively uniseriate, 290-721-1350 µm high, 46-50-55 ray/mm.

Anatomical differences to stem xylem: Mean tangential diameter of vessel lumina is similar in the branch and in the stem (10-29-40/20-35-40 µm) while vessel density is lower in the branch (105-125-140/129-155-180 vessels/mm²). Rays are higher in the branch (290-721-1350/100-306-510 µm) while ray number is similar (46-50-55/23-26-30 rays/mm).

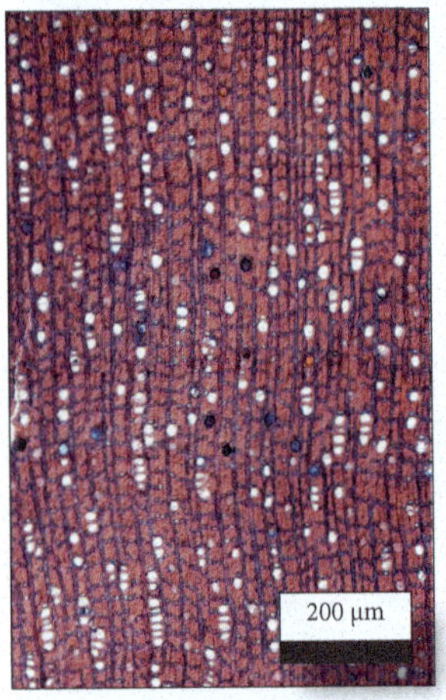

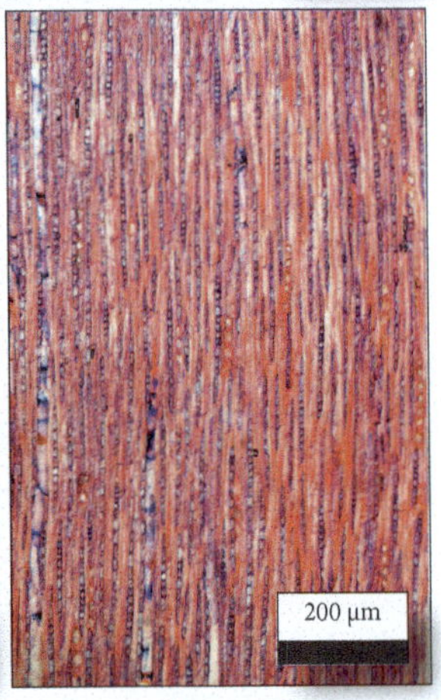

BARK ANATOMY

Branch bark: Collapsed and non-collapsed sieve tubes present in the phloem. Groups of sclereids scattered dispersed present. The cortex consists of alternating bands of sclereids and irregular rectangular blue-stained parenchyma cells.

Stem bark: Phloem consists of alternating groups of tangentially distributed sclereids, and collapsed and non-collapsed sieve tubes. Prismatic crystals present in the sclereids and scattered or irregularly dispersed in the phloem. The phellem is homogeneous, it consists of regularly arranged rectangular cells. Lignified cells in phellem, distinct in polarized light.

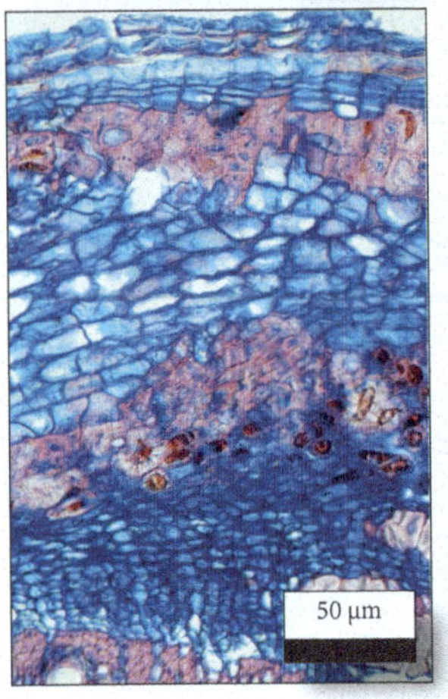

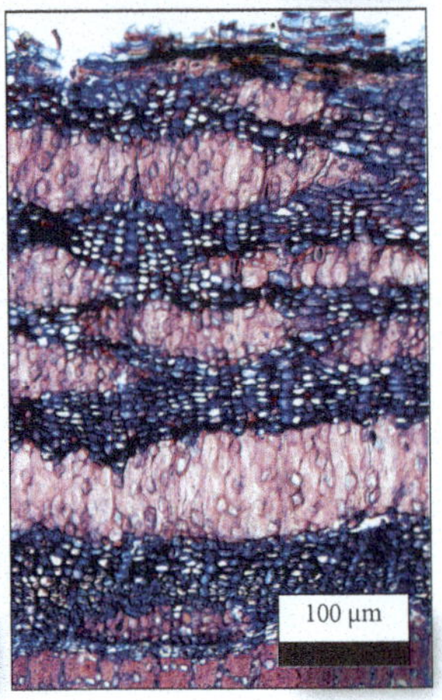

IAWA feature numbers
2 5 9.1 10 13 22 25 30 40 50 52 58 61 66 70 77 78 92 93 96 102.3 107 116 136 142 180 189 192 196

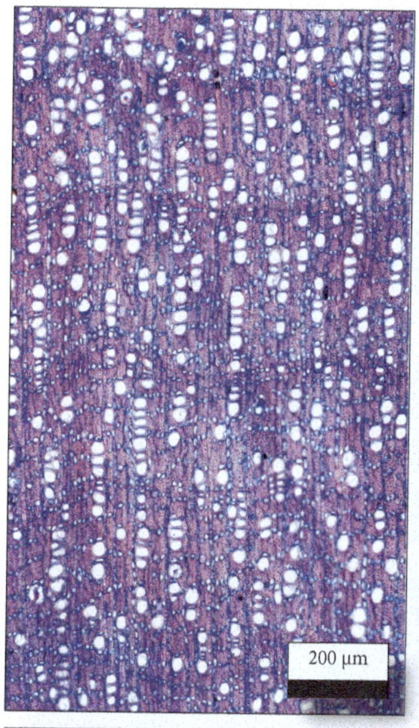

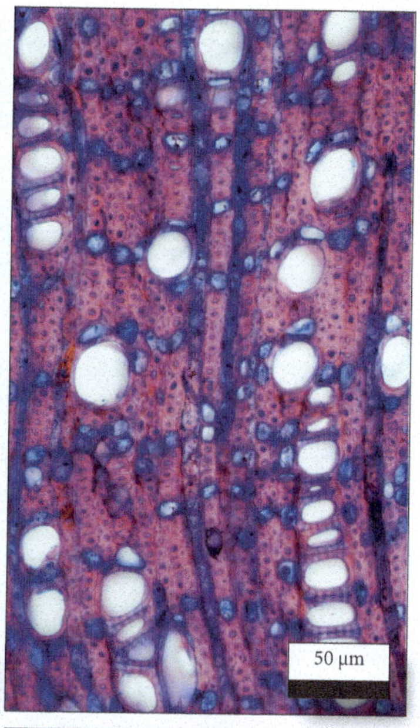

STEM XYLEM ANATOMY

Transverse section

Growth ring boundaries absent.

Wood diffuse-porous.

Vessels solitary and in short to long radial multiples of 2 to 6(10) elements.

Tangential diameter of vessels 20-35-40 µm, 129-155-180 vessels/mm^2.

Gums and/or other deposits in heartwood vessels.

Fibers very thick-walled.

Apotracheal parenchyma diffuse-in-aggregates.

Paratracheal parenchyma scanty.

Rays 23-26-30/mm.

Radial section

Simple perforation plates.

Intervessel pits circular and alternate, 4-7 µm in diameter.

Vessel-ray pits with distinct borders; similar to intervessel pits in size and shape throughout the ray cell.

Vessel element length 150-193-240 µm.

Fibres with simple to minutely bordered pits (libriform fibers).

Heterogeneous rays, body ray cells procumbent with 2 to 4 rows of upright and square marginal cells.

Prismatic crystal in chambered axial parenchyma cells.

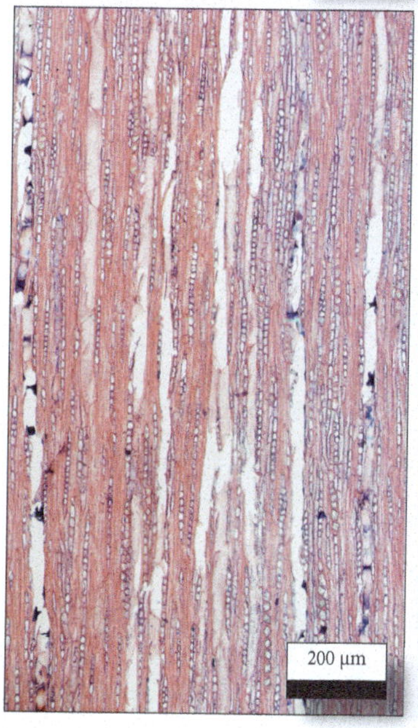

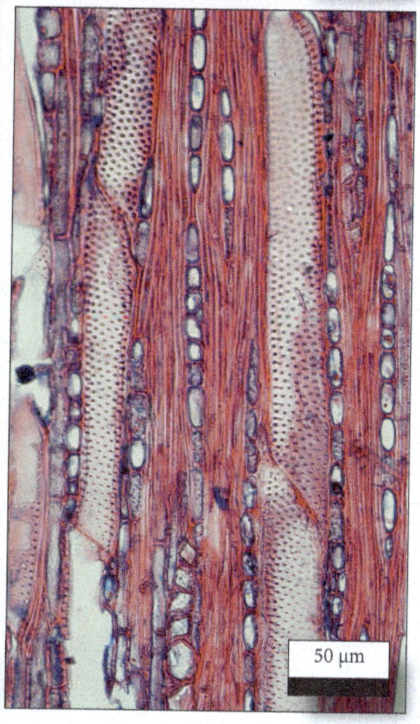

Tangential section

Rays exclusively uniseriate.

Ray height 100-306-510 µm.

Three to 8 cells per parenchyma strand.

Diospyros ankifiensis H. Perrier.

EN: Ebony FR: Bois d'ébène ML: Hazomainty

PLANT DESCRIPTION

Shrub to tree | 6-10 m high
Endemic to Madagascar. Non-exploited species.

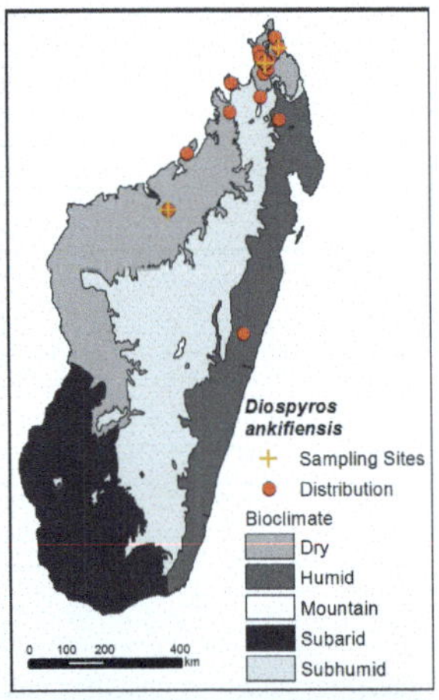

DISTRIBUTION

Elevation: 0-500 m a.s.l.
Vegetation formation: Forest
Bioclimate: Dry and sub-humid
Protected areas: Ankarana, Ankarafantsika and Lokobe
New protected areas: Sahafary

Slides made from material collected in Ankarana.

PITH AND PRIMARY XYLEM ANATOMY

Pith: Shape of pith round. Cells dimorphic. Prismatic crystals present. Pits in transverse cell walls. Pith not visible in polarized light.

Primary xylem: Vascular bundles separated. Vessels solitary and in short radial multiples.

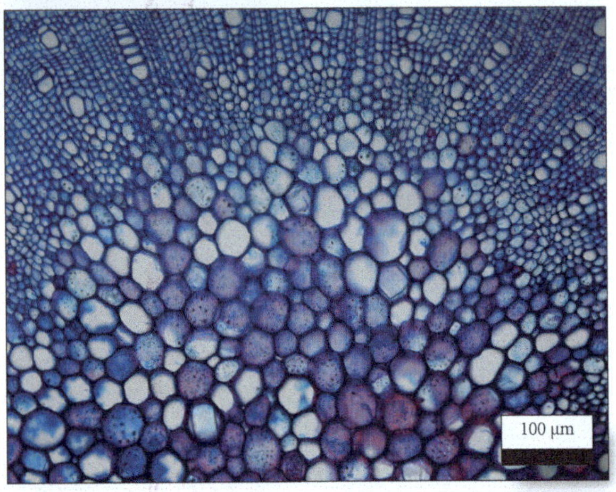

BRANCH XYLEM ANATOMY

Description: Growth rings boundaries indistinct or absent. Tangential diameter of vessels 40-58-90 µm, 28-35-42 vessels/mm^2. Vessel element length 200-332-420 µm. Intervessel pits small, 4-7 µm in diameter. Fibres very thick-walled. Axial parenchyma diffuse-in-aggregates. Rays exclusively uniseriate, 160-494-800 µm high. Rays 30-34-38/mm.

Anatomical differences to stem xylem: Mean tangential diameter of vessel lumina is slightly lower in the branch (40-58-90/60-75-90 µm). Vessel density is similar (28-35-42/30-39-49 vessels/mm^2). Rays are higher in the branch (160-494-800/150-413-800 µm). Ray number is higher in the branch (30-34-38/22-27-31 rays/mm).

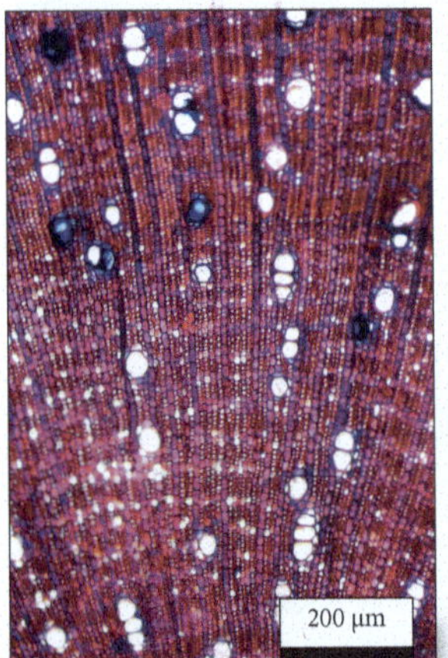

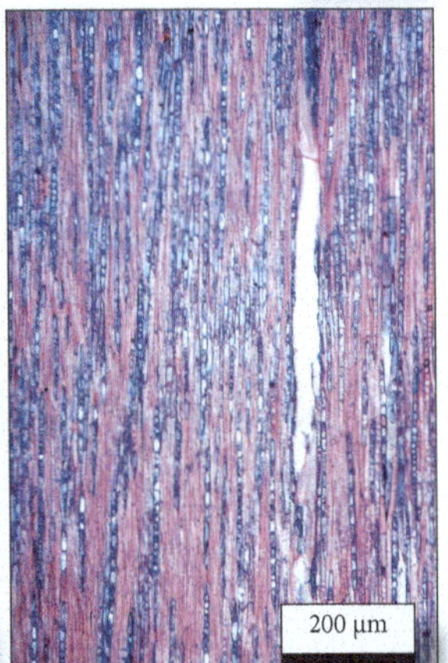

BARK ANATOMY

Branch bark: Collapsed and non-collapsed sieve tubes present in the phloem. Prismatic crystals present. The cortex is characterized by a band of sclereids and irregularly blue-stained parenchyma cells.

Stem bark: The phloem is composed by collapsed and non-collapsed sieve tubes, and tangentially distributed groups of sclereids. Ray dilatation present. The cortex is presented by a band of sclereids. The phellem is homogeneous with dark staining substances and it consists of regularly arranged rectangular cells, indistinct in polarized light.

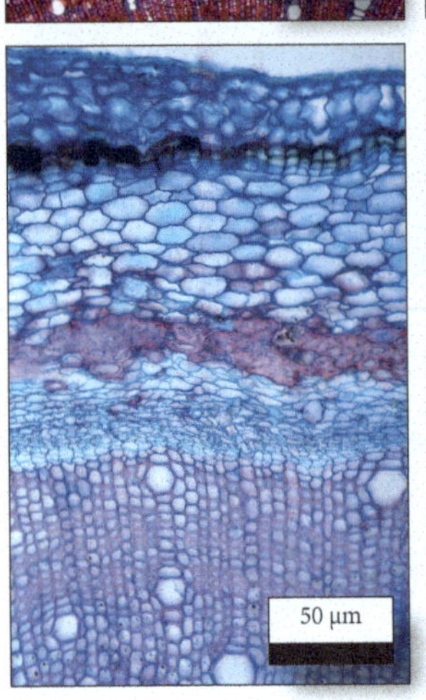

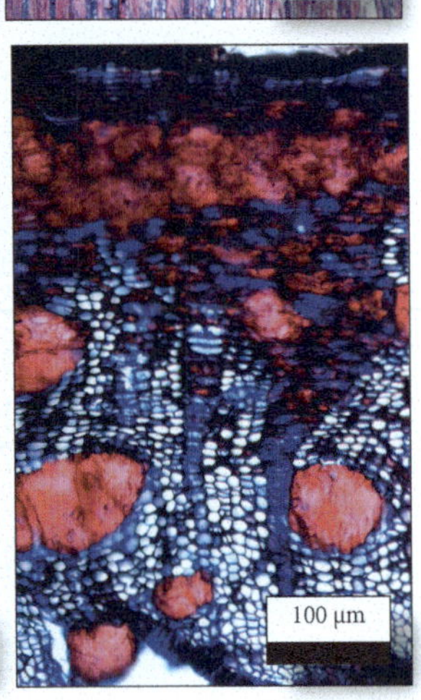

IAWA feature numbers
2 5 9.1 10 13 22 25 30 41 48 52 58 61 66 70 77 78 92 96
102.3 107 108 116 136 142 180 189 190 192.2 196

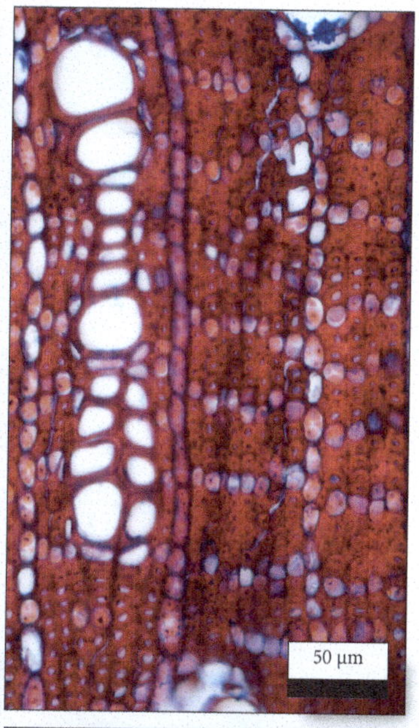

STEM XYLEM ANATOMY

Transverse section
Growth ring boundaries absent.
Wood diffuse-porous.
Vessels solitary and in radial multiples of 2-5(10) elements.
Tangential diameter of vessels 60-75-90 µm, 30-39-49 vessels/mm^2.
Gums and/or other deposits in heartwood vessels.
Fibers very thick-walled.
Apotracheal parenchyma diffuse-in-aggregates.
Paratracheal parenchyma scanty.
Rays 22-27-31/mm.

Radial section
Simple perforation plates.
Intervessel pits circular and alternate, 4-7 µm in diameter.
Vessel-ray pits with distinct borders; similar to intervessel pits in size and shape throughout the ray cell.
Vessel element length 270-315-380 µm.
Fibres with simple to minutely bordered pits (libriform fibers).
Heterogeneous rays, body ray cells procumbent with 2 to over 4 rows of upright and square marginal cells.
Prismatic crystal in chambered axial parenchyma cells.

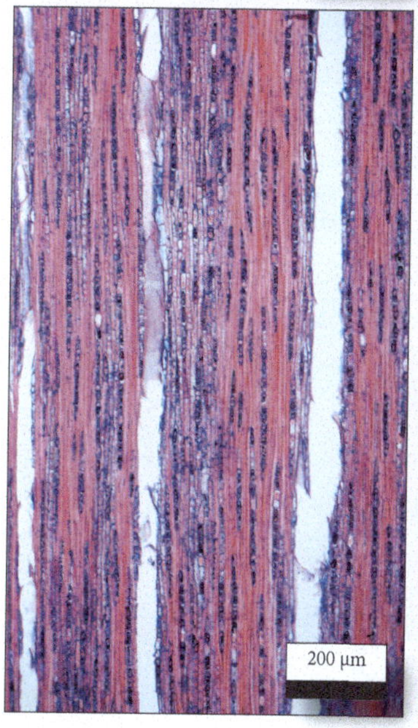

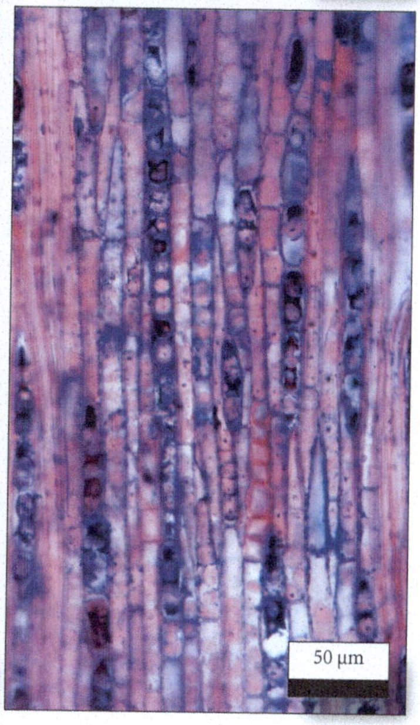

Tangential section
Ray exclusively uniseriate.
Ray height 150-413-800 µm.
Two to 4 cells per parenchyma strand.

Diospyros antongiliensis ined.

EN: Ebony FR: Bois d'ébène ML: Hazomainty

PLANT DESCRIPTION

Tree | 5-10 m high | 6-18 cm DBH
Endemic to Madagascar. Non-exploited species.

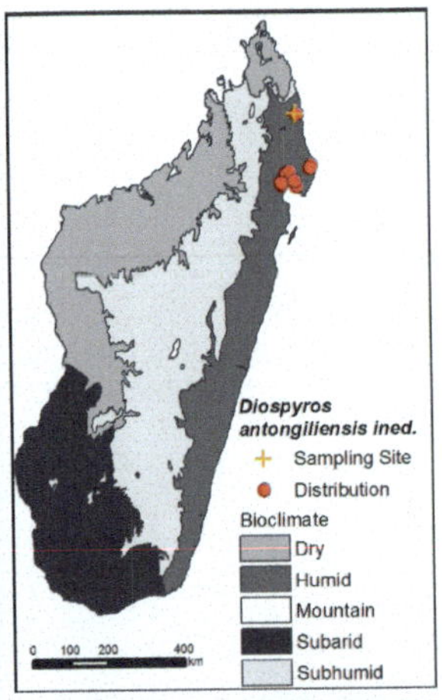

DISTRIBUTION

Elevation: 0-500 m a.s.l.
Vegetation formation: Forest
Bioclimate: Humid
Protected areas: -
New protected areas: Makirovana

Slides made from material collected in Makirovana.

PITH AND PRIMARY XYLEM ANATOMY

Pith: Shape of pith round. Thick-walled parenchyma cells present. Cells dimorphic. Prismatic crystals present. Pits in transverse cell walls. Vascular bundles separated. Pith not visible in polarized light.

Primary xylem: Vascular bundles separated. Vessels solitary and in long radial multiples.

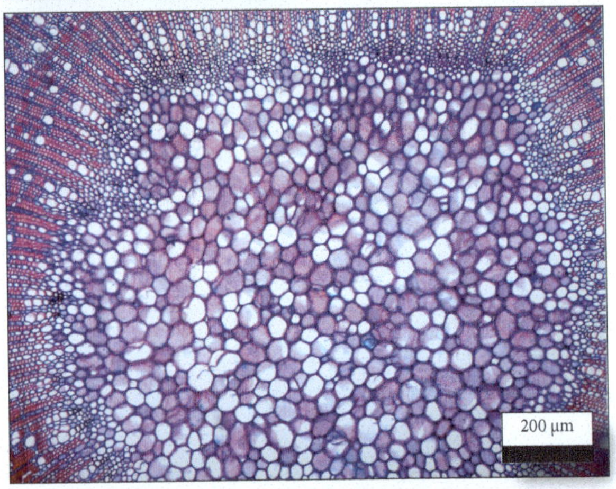

BRANCH XYLEM ANATOMY

Description: Growth rings boundaries indistinct or absent. Tangential diameter of vessels 70-82-100 µm, 8-18-29 vessels/mm^2. Vessel element length 230-365-500 µm. Intervessel pits small, 4-7 µm in diameter. Fibers thick- to very thick-walled. Axial parenchyma diffuse-in-aggregates. Rays exclusively uniseriate. Ray height 130-297-450 µm. Rays 27-30-35/mm.

Anatomical differences to stem xylem: Mean tangential diameter of vessel lumina is similar in the branch and in the stem (70-82-100/60-84-100 µm), while vessel density is slightly higher in the branch (8-18-29/11-16-18 vessels/mm^2). Rays are shorter in branch (130-297-450/330-426-560 µm), and more numerous (27-30-35/18-20-23 rays/mm).

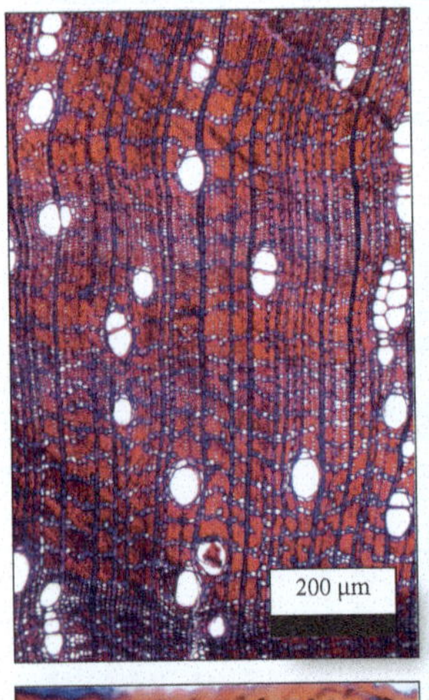

BARK ANATOMY

Branch bark: Collapsed and non-collapsed sieve tubes present in the phloem. The cortex is characterized by a band of sclereids and irregular oval blue-stained parenchyma cells.

Stem bark: The phloem is characterized by collapsed and non-collapsed sieve tubes, and irregularly distributed groups of sclereids. Ray dilatation present. The phellem is homogeneous with dark stained substances and it consists of regularly arranged rectangular cells, indistinct in polarized light.

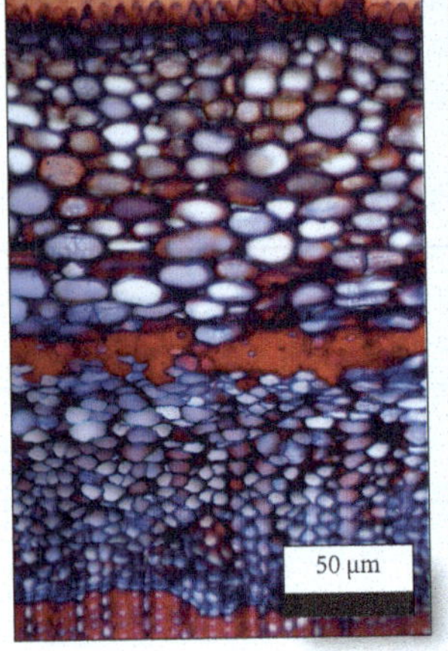

IAWA feature numbers
2 5 9 9.1 13 22 25 30 41 47 53 58 61 66 70 77 86 92 93
96 102.3 107 116 136 142 180 189 192 196

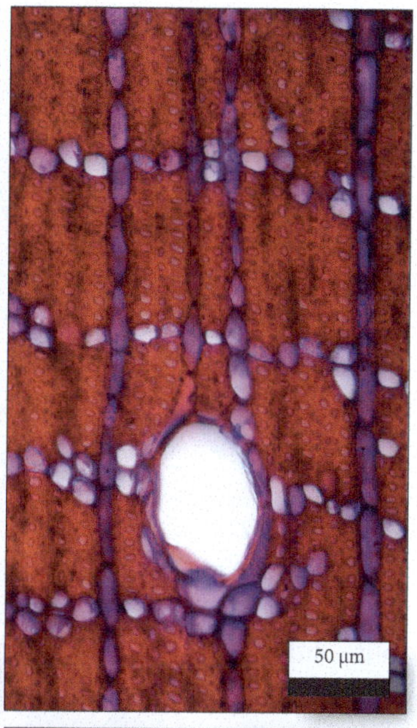

STEM XYLEM ANATOMY

Transverse section

Growth rings boundaries indistinct or absent.
Wood diffuse-porous.
Vessels solitary and in short radial multiples of 2 to 4 elements.
Tangential diameter of vessels medium 60-84-100 µm, 11-16-18 vessels/mm^2.
Gums and/or other deposits in heartwood vessels.
Fibers very thick-walled.
Apotracheal parenchyma diffuse-in-aggregates and in narrow lines up to 3 cells wide. Paratracheal parenchyma scanty.
Rays 18-20-23/mm.

Radial section

Simple perforation plates.
Intervessel pits circular and alternate, 4-7 µm in diameter.
Vessel-ray pits with distinct borders; similar to intervessel pits in size and shape throughout the ray cell.
Vessel element length 330-426-560 µm.
Fibres with simple to minutely bordered pits (libriform fibers).
Heterogeneous rays, body ray cells procumbent with 1 to 4 rows of upright and square marginal cells.
Prismatic crystal in chambered axial parenchyma cells.

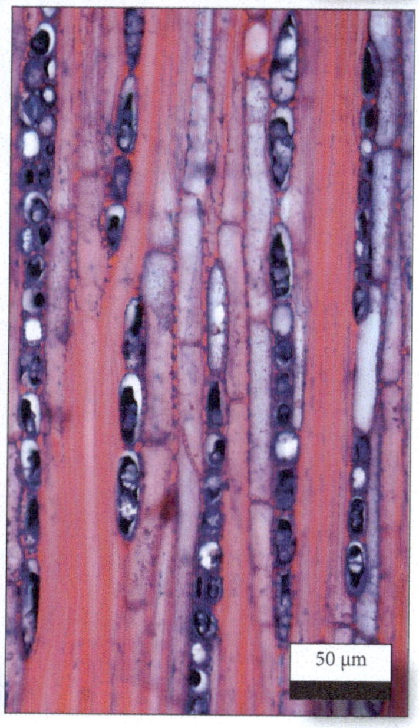

Tangential section

Rays exclusively uniseriate.
Ray height 120-447-810 µm.
Dark staining substances in heartwood ray cells.
Three to 8 cells per parenchyma strand.

Diospyros bernieriana (Baill.) H. perrier

EN: Ebony FR: Bois d'ébène ML: Hazomainty

PLANT DESCRIPTION

Tree | 4-15 m high | 8 cm DBH
Endemic to Madagascar. Non-exploited species.

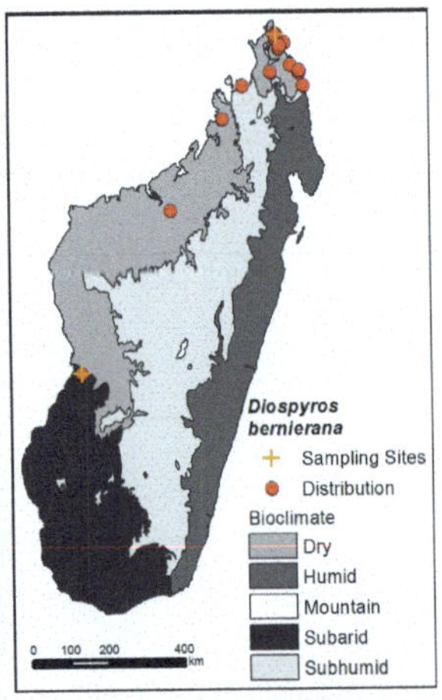

DISTRIBUTION

Elevation: 0-500 m a.s.l.
Vegetation formation: Forest
Bioclimate: Humid
Protected areas: Bemaraha and Lokobe
New protected areas: Sahafary

Slides made from material collected in Sahafary.

PITH AND PRIMARY XYLEM ANATOMY

Pith: Shape of pith round. Thick-walled parenchyma cells present. Cells dimorphic. Pits in transverse cell walls. Pith visible in polarized light.

Primary xylem: Vascular bundles separated. Vessels solitary and in long radial multiples.

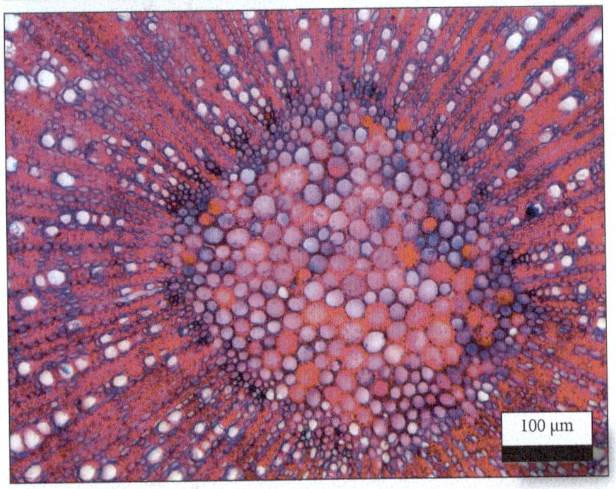

BRANCH XYLEM ANATOMY

Description: Growth rings boundaries indistinct or absent. Tangential diameter of vessels 10-25-40 µm, 120-140-160 vessels/mm². Vessel element length 260-337-460 µm. Intervessel pits small, 4-7 µm in diameter. Fibres very thick-walled. Axial parenchyma diffuse-in-aggregates. Rays exclusively uniseriate. Ray height 290-536-1300 µm. Rays 43-48-52/mm.

Anatomical differences to stem xylem: Mean tangential diameter of vessel lumina is slightly smaller in the branch (10-25-40/20-29-40 µm). Vessel density is higher in branch (120-140-160/85-92-102 vessels/mm²). Ray height is higher in the branch (290-536-1300/90-332-410 µm), while ray number is slightly higher in the stem (43-48-52/27-30-34 rays/mm).

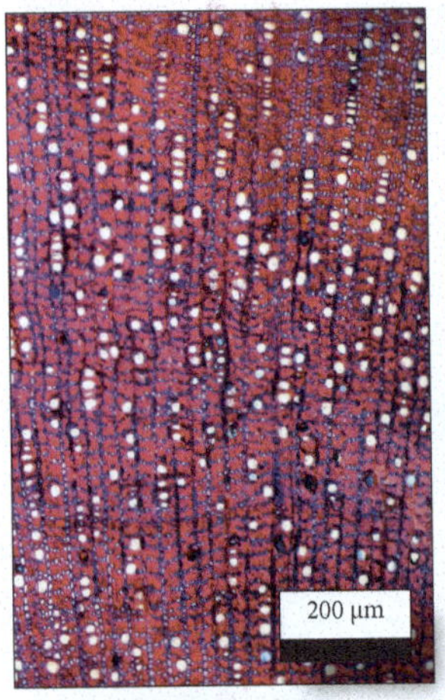

BARK ANATOMY

Branch bark: Collapsed and non-collapsed sieve tubes present in the phloem. The cortex is characterized by a band of sclereids and irregularly blue-stained parenchyma cells.

Stem bark: Tangentially distributed solitary and groups of sclereids, and collapsed and non-collapsed sieve tubes present in the phloem. Ray and Axial parenchyma dilatation present. Band of sclereids present in the cortex. The phellem is homogeneous with dark stained substances and it consists of regularly arranged rectangular cells.

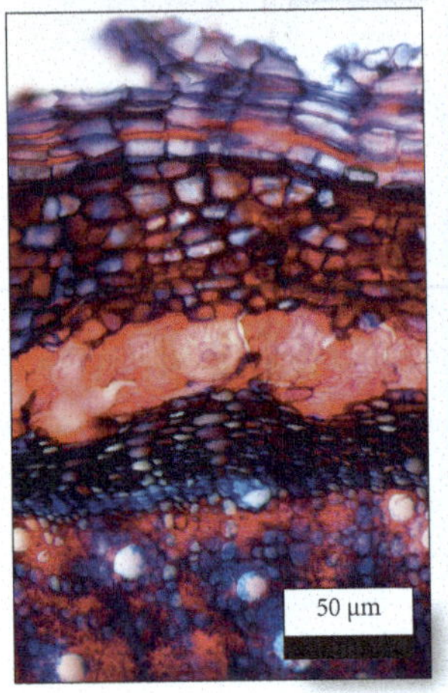

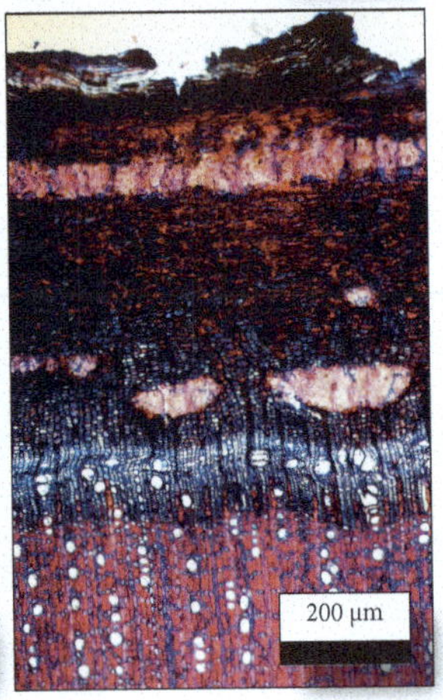

IAWA feature numbers
2 5 9 9.1 13 22 25 30 40 49 52 58 61 66 70 77 78 92 96.1
102.4 108 116 136 142 180 189 192 196

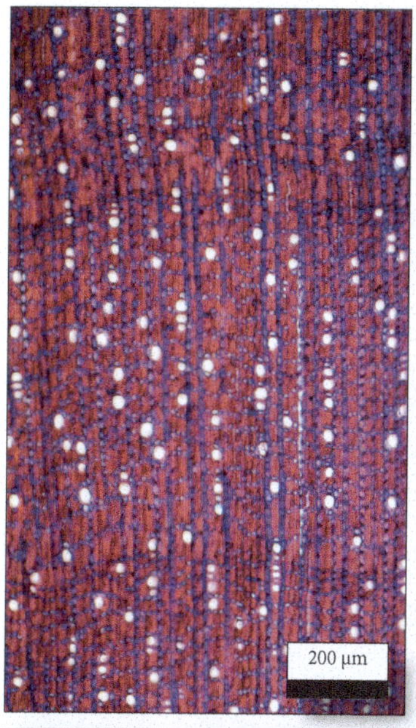

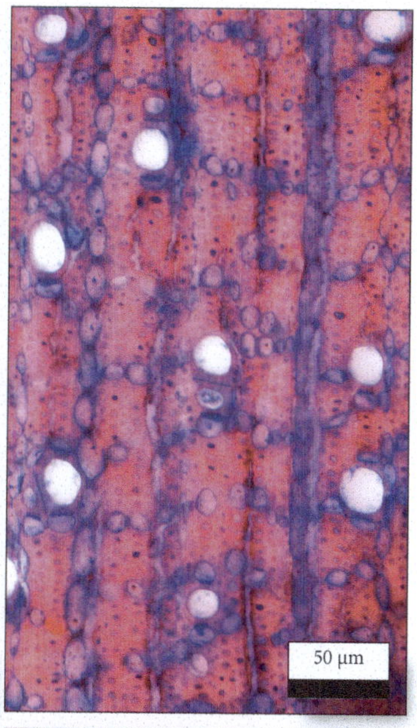

STEM XYLEM ANATOMY

Transverse section
Growth ring boundaries absent.
Wood diffuse-porous.
Vessels solitary and in short radial multiples of 2 to 4 elements.
Tangential diameter of vessels 20-29-40 µm, 85-92-102 vessels/mm^2.
Gums and/or other deposits in heartwood vessels.
Fibers very thick-walled.
Apotracheal parenchyma diffuse-in-aggregates, paratracheal parenchyma scanty.
Rays 27-30-34/mm.

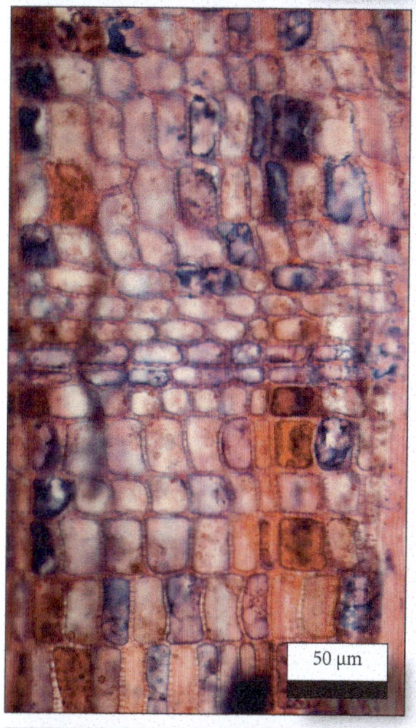

Radial section
Simple perforation plates.
Intervessel pits circular and alternate, 4-7 µm in diameter.
Vessel-ray pits with distinct borders; similar to intervessel pits in size and shape throughout the ray cell.
Vessel element length 230-332-410 µm.
Fibres with simple to minutely bordered pits (libriform fibers).
Heterogeneous rays, body ray cells procumbent with over 4 rows of upright and square marginal cells.
Prismatic crystal in chambered axial parenchyma cells and in procumbent ray cells.

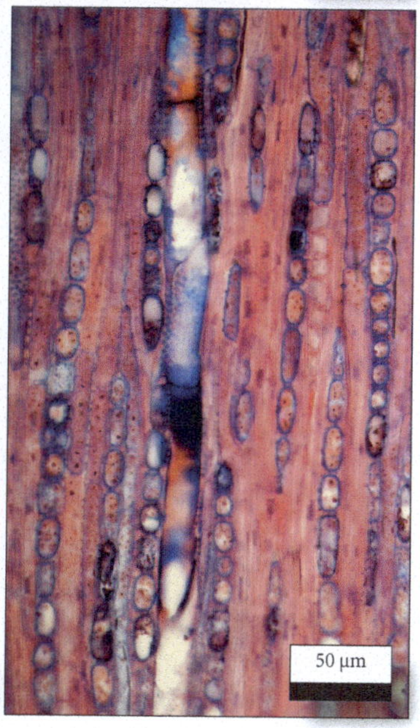

Tangential section
Rays uni- and biseriate, predominantly uniseriate.
Ray height 90-332-410 µm.
Two to 4 cells per parenchyma strand.

Diospyros brevipedicellata ined.

EN: Ebony   FR: Bois d'ébène   ML: Hazomainty

PLANT DESCRIPTION

Tree | 4-10 m high | 6-18 cm DBH
Endemic to Madagascar. Non-exploited species.

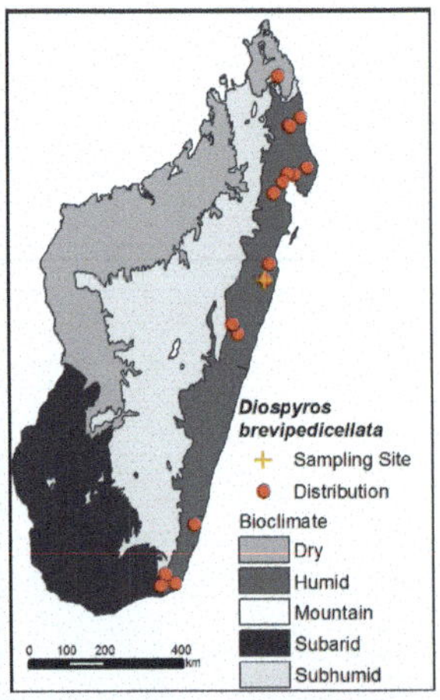

DISTRIBUTION

Elevation: 0-1000 m a.s.l.
Vegetation formation: Forest
Bioclimate: Humid
Protected areas: Betampona, Andohahela, Mantadia and Nosy Mangabe
New protected areas: Mahabo-Mananivo and Daraina

Slides made from material collected in Betampona.

PITH AND PRIMARY XYLEM ANATOMY

Pith: Shape of pith round. Thick-walled parenchyma cells present. Cells dimorphic. Prismatic crystals present. Pits in transverse cell walls. Pith visible in polarized light.

Primary xylem: Vascular bundles separated. Vessels solitary and in long radial multiples.

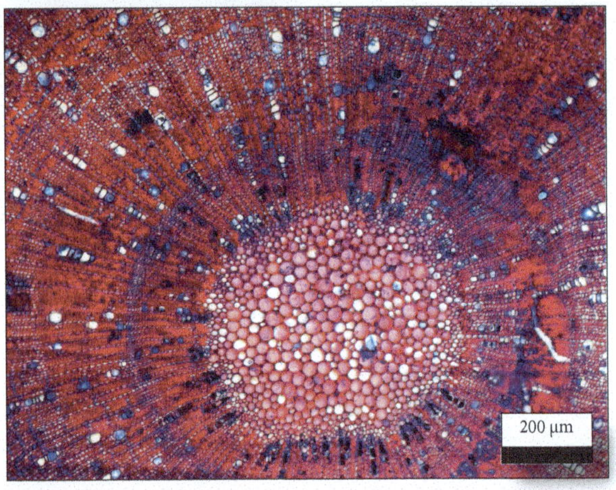

BRANCH XYLEM ANATOMY

Description: Growth rings boundaries indistinct or absent. Tangential diameter of vessels 50-71-90 µm, 30-40-48 vessels/mm². Vessel element length 300-418-620 µm. Intervessel pits small, 4-7 µm in diameter. Fibres thin- to thick-walled. Axial parenchyma diffuse-in-aggregates. Rays exclusively uniseriate. Ray height 150-382-700 µm. Rays 31-35-39/mm.

Anatomical differences to stem xylem: Mean tangential diameter of vessel lumina is smaller in the branch (50-71-90/90-104-140 µm). Vessel density is higher in the branch (30-40-48/22-29-42 vessels/mm²). Rays height and number are smaller in the branch (150-382-700/160-588-1320 µm; 31-35-39/12-15-19 rays/mm).

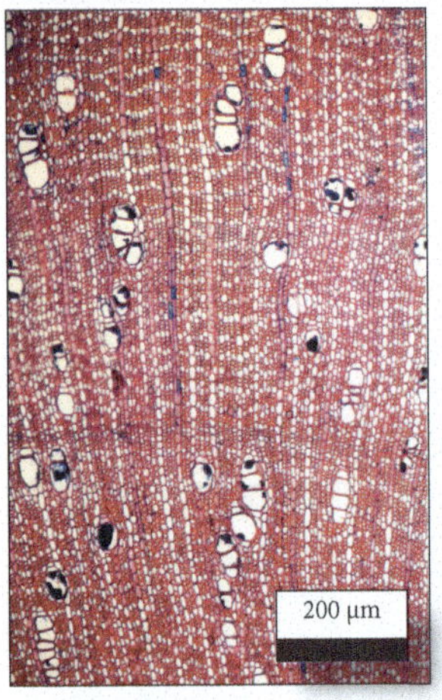

BARK ANATOMY

Branch bark: Collapsed and non-collapsed sieve tubes present in the phloem. The cortex is characterized by a band of sclereids and irregularly shaped blue-stained parenchyma cells.

Stem bark: The phloem is characterized by collapsed and non-collapsed sieve tubes and groups of fibres in radial rows. Groups of sclereids dispersed present. Ray and Axial parenchyma dilatation present. The phellem is homogeneous with dark stained substances and it consists of regularly arranged rectangular cells, indistinct in polarized light.

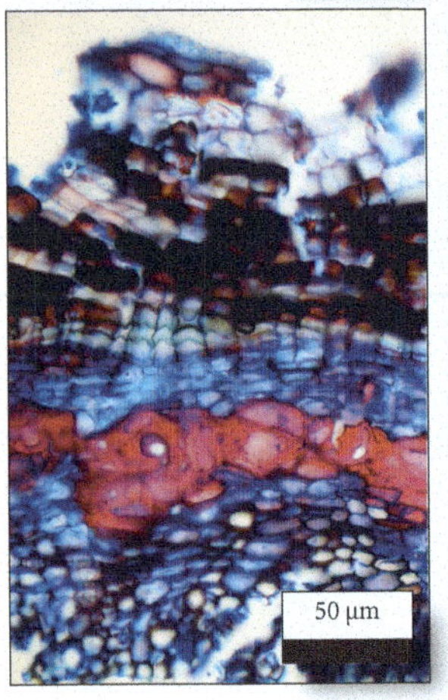

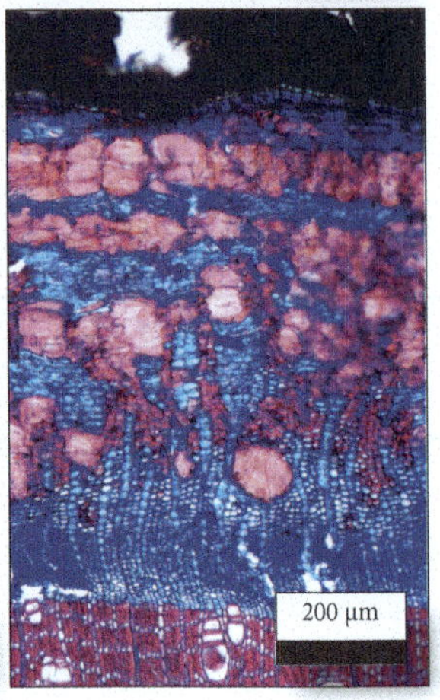

IAWA feature numbers
2 5 10 13 22 25 30 42 48 53 58 61 66 70 77 79 86 92 96.1
102.4 108 109 116 136 138 142 180 189 192 196

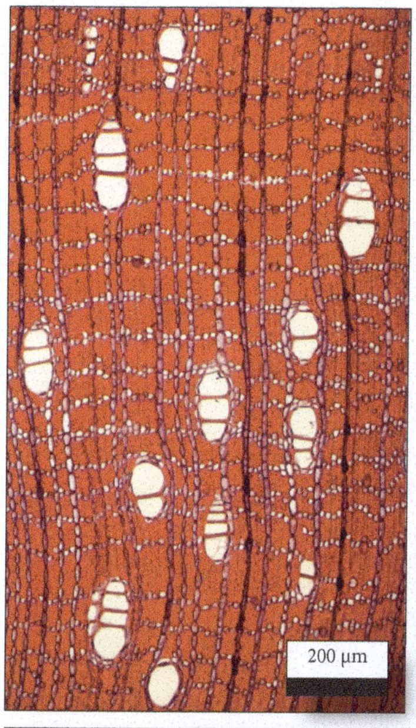

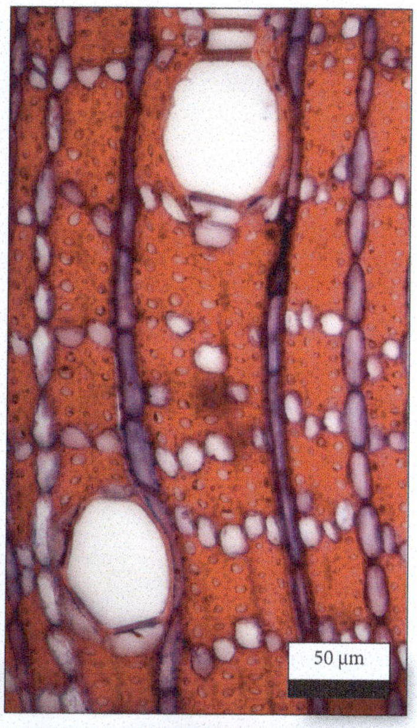

STEM XYLEM ANATOMY

Transverse section
Growth ring boundaries absent.
Wood diffuse-porous.
Vessels in radial multiples of 2-5 elements.
Tangential diameter of vessels 90-104-140 µm, 22-29-42 vessels/mm^2.
Gums and/or other deposits in heartwood vessels.
Fibers very thick-walled.
Apotracheal parenchyma diffuse-in-aggregates and in narrow lines up to 3 cells, paratracheal parenchyma scanty.
Rays 12-15-19/mm.

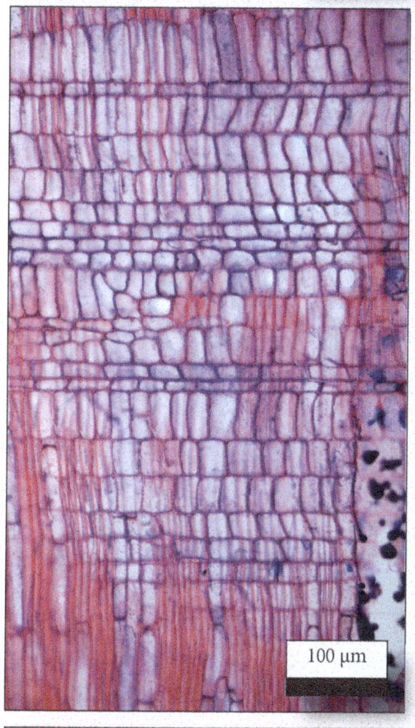

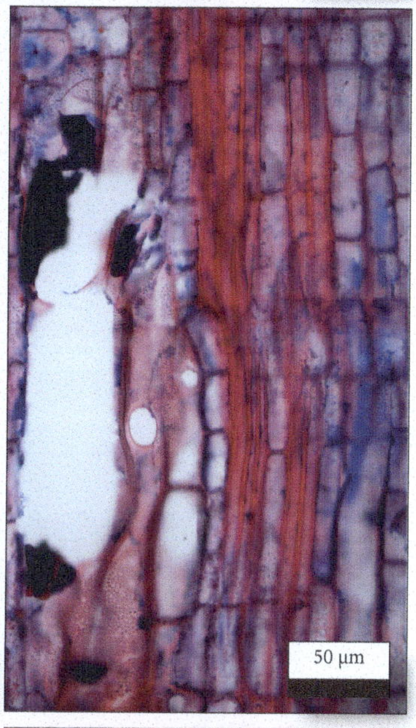

Radial section
Simple perforation plates.
Intervessel pits circular and alternate, 4-7 µm in diameter.
Vessel-ray pits with distinct borders; similar to intervessel pits in size and shape throughout the ray cell.
Vessel element length 270-425-750 µm.
Fibres with simple to minutely bordered pits (libriform fibers).
Heterogeneous rays, body ray cells procumbent with over 4 rows of upright and square marginal cells. Some rays with procumbent, square and upright cells mixed throughout the ray.
Prismatic crystal in chambered axial parenchyma cells and in procumbent ray cells.

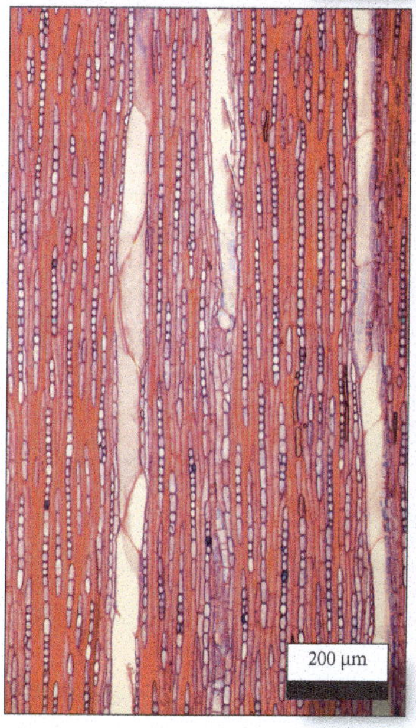

Tangential section
Rays uni- and biseriate, predominantly uniseriate.
Ray height 160-588-1320 µm.
Two to 4 cells per parenchyma strand.

Diospyros calophylla Hiern.

EN: Ebony FR: Bois d'ébène ML: Hazomainty

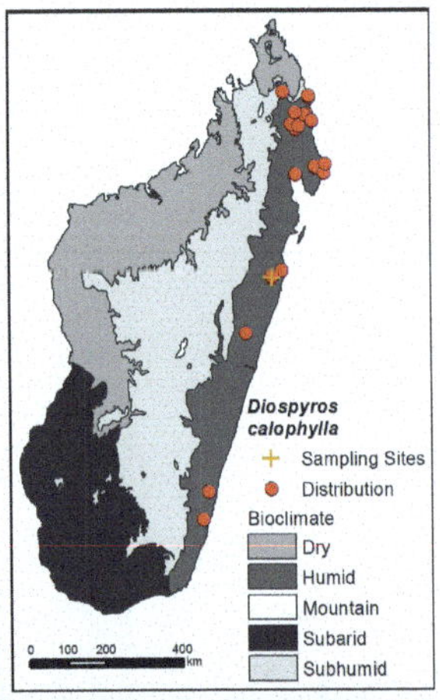

PLANT DESCRIPTION

Shrub to tree | 2-25 m high | 4-32 cm DBH
Endemic to Madagascar. Exploited species.

DISTRIBUTION

Elevation: 0-1000 m a.s.l.
Vegetation formation: Forest
Bioclimate: Humid
Protected areas: Betampona, Marojejy, Masoala and Nosy Mangabe
New protected areas: Analalava

Slides made from material collected in Analalava.

PITH AND PRIMARY XYLEM ANATOMY

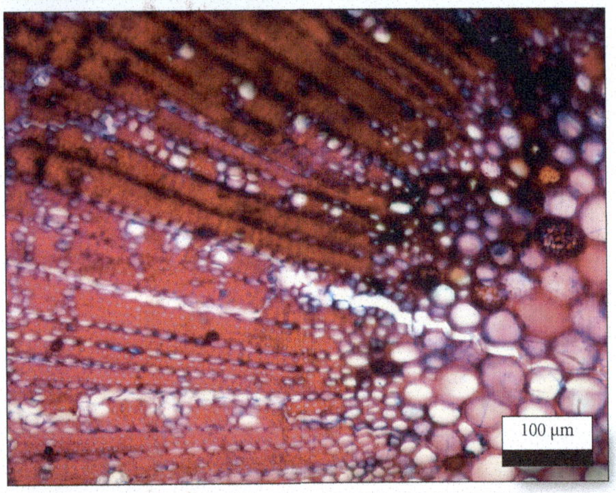

Pith: Shape of pith round. Thick-walled parenchyma cells present. Cells dimorphic. Pits in transverse cell walls. Pith not visible in polarized light.

Primary xylem: Vascular bundles not distinct. Vessels solitary and in short radial multiples.

BRANCH XYLEM ANATOMY

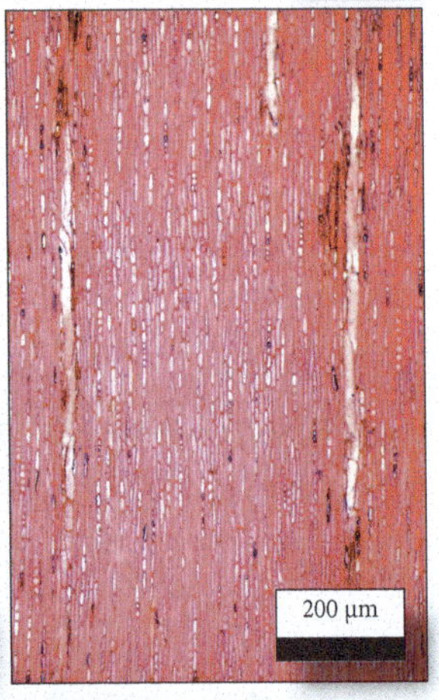

Description: Growth rings boundaries indistinct or absent. Tangential diameter of vessels 10-24-40 µm, 23-41-50 vessels/mm^2. Vessel element length 280-374-590 µm. Intervessel pits small, 4-7 µm in diameter. Fibers thick- to very thick-walled. Apotracheal parenchyma diffuse-in-aggregates, paratracheal parenchyma scanty to vasicentric. Rays exclusively uniseriate. Ray height 250-511-950 µm. Rays 38-42-48/mm.

Anatomical differences to stem xylem: Mean tangential diameter of vessel lumina is similar in the branch and in the stem (10-24-40/12-28-48 µm) while vessel density is higher in the branch than in the stem (23-41-50/16-25-38 vessels/mm^2). Rays are higher and more frequent in the branch (250-511-950/160-416-710 µm; 38-42-48/24-27-29 rays/mm).

BARK ANATOMY

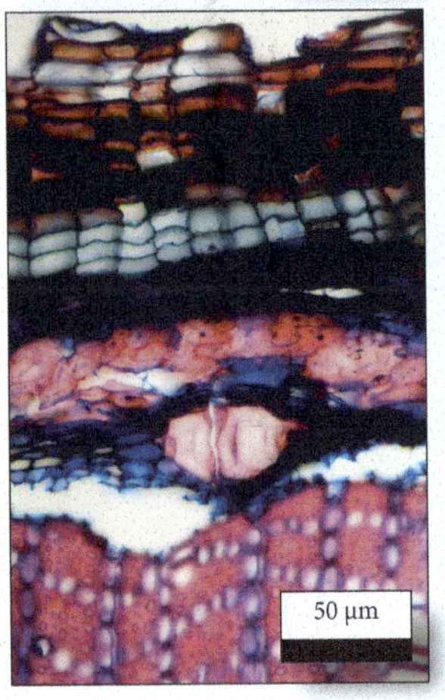

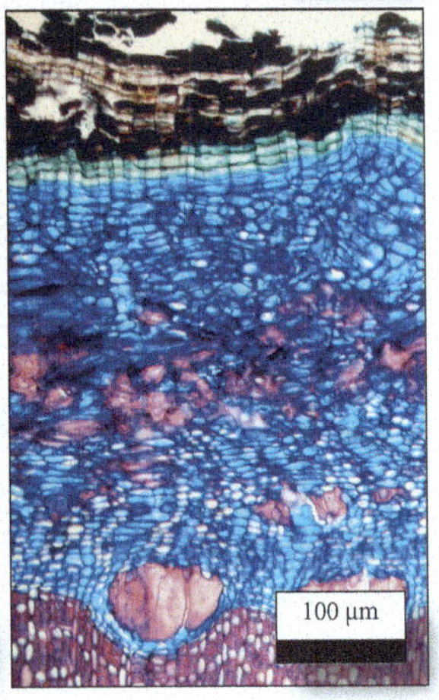

Branch bark: The phloem is characterized by collapsed and non-collapsed sieve tubes and groups of sclereids irregularly dispersed. The cortex is characterized by a band of sclereids and irregular dark stained parenchyma cells.

Stem bark: The phloem consists of collapsed and non-collapsed sieve tubes, and irregularly distributed groups of sclereids. Some rays become dilated. The phellem is homogeneous with dark stained substances and it consists of regularly arranged rectangular cells, indistinct in polarized light.

IAWA feature numbers
2 5 9 9.1 13 22 25 30 40 48 53 58 61 66 69 70 77 78 79
92 96 102.3 107 108 116 136 138 142 180 189 192 196

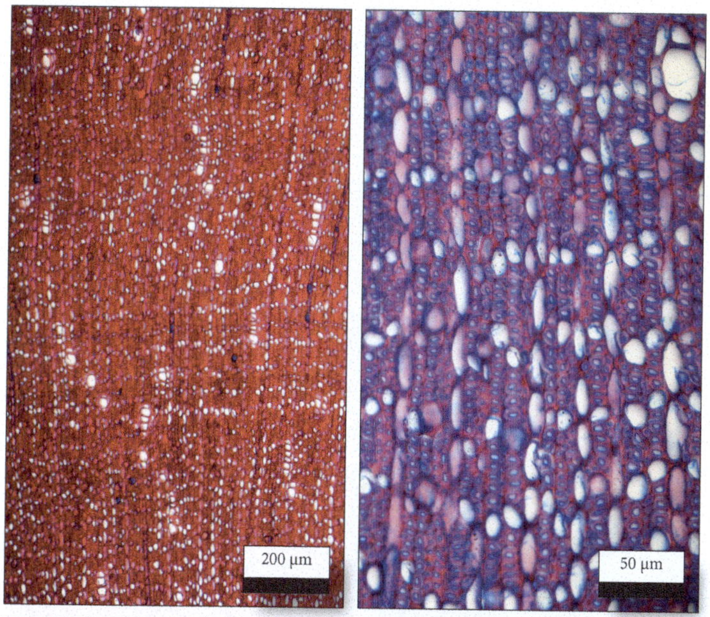

STEM XYLEM ANATOMY

Transverse section

Growth rings boundaries indistinct or absent.
Wood diffuse-porous.
Vessels solitary and in radial multiples of 2 to 5 elements.
Tangential diameter of vessels small 12-28-48 μm, 16-25-38 vessels/mm^2.
Gums and/or other deposits in heartwood vessels.
Fibers thick- to very thick-walled.
Apotracheal parenchyma diffuse-in-aggregates, paratracheal parenchyma scanty to vasicentric.
Rays 24-27-29/mm.

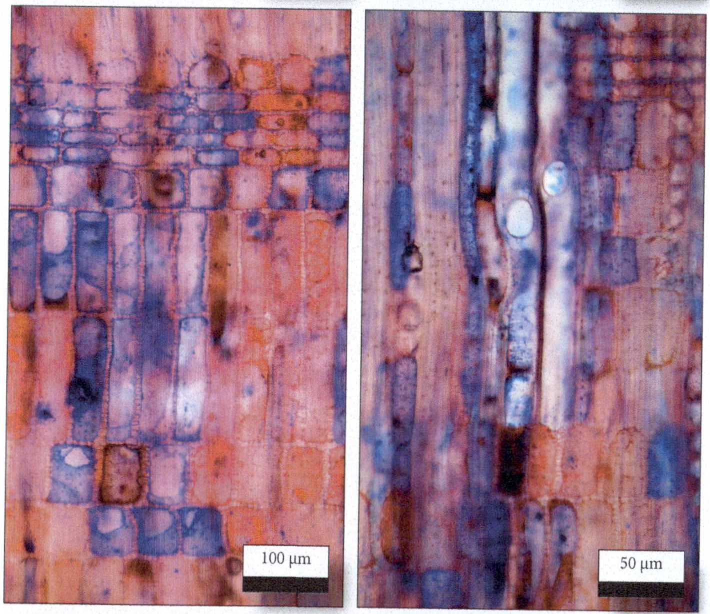

Radial section

Simple perforation plates.
Intervessel pits circular and alternate, 4-7 μm in diameter.
Vessel-ray pits with distinct borders; similar to intervessel pits in size and shape throughout the ray cell.
Vessel element length 220-359-460 μm.
Fibres with simple to minutely bordered pits (libriform fibers).
Heterogeneous rays, body ray cells procumbent with 2 to over 4 rows of upright and square marginal cells.
Prismatic crystal in chambered axial parenchyma cells, and in procumbent ray cells.

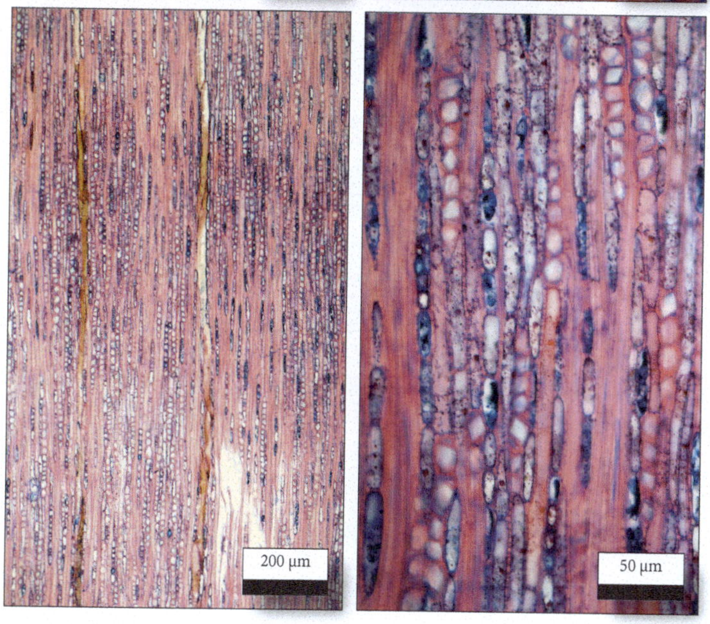

Tangential section

Rays exclusively uniseriate.
Ray height 160-416-710 μm.
Two to 4 cells per parenchyma strand.

Diospyros chitinophora ined.

EN: Ebony FR: Bois d'ébène ML: Hazomainty

PLANT DESCRIPTION

Tree | 5-15 m high | 18 cm DBH
Endemic to Madagascar. Non-exploited species.

DISTRIBUTION

Elevation: 0-500 m a.s.l.
Vegetation formation: Forest
Bioclimate: Dry
Protected areas: Ankarana

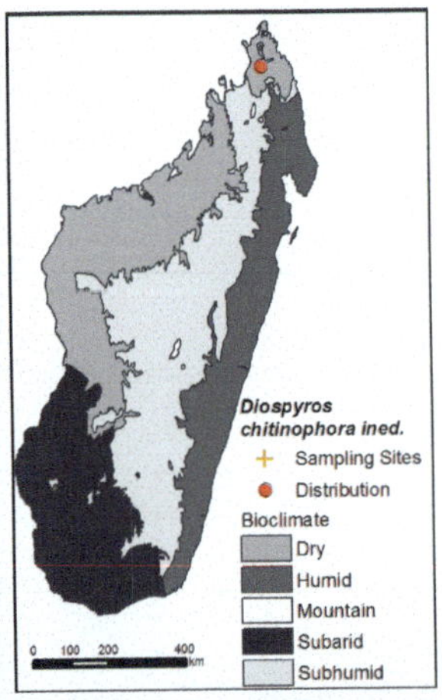

Slides made from material collected in Ankarana

PITH AND PRIMARY XYLEM ANATOMY

Pith: Shape of pith oval. Thick-walled parenchyma cells present. Cells dimorphic. Prismatic crystals present. Pits in transverse cell walls. Pith not visible in polarized light.

Primary xylem: Vascular bundles separated. Vessels solitary and in long radial multiples.

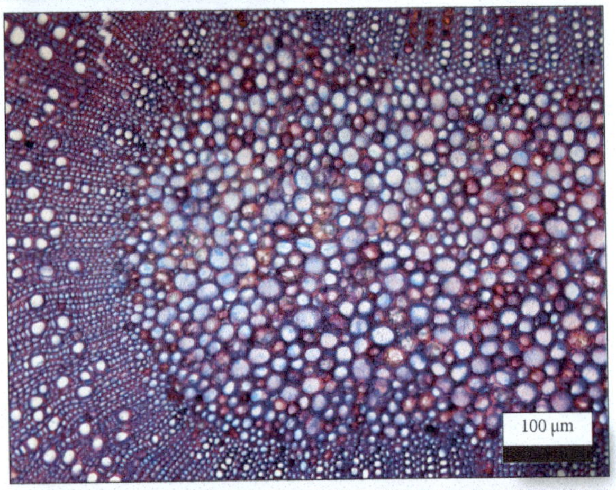

BRANCH XYLEM ANATOMY

Description: Growth rings boundaries indistinct or absent. Tangential diameter of vessels 7-12-15 µm, 111-131-146 vessels/mm^2. Vessel element length 220-322-450 µm. Intervessel pits small, 4-7 µm in diameter. Fibers thick- to very thick-walled. Axial parenchyma diffuse and diffuse-in-aggregates. Rays uni- and biseriate, predominantly uniseriate. Ray height 140-254-370 µm, 38-41-44 rays/mm.

Anatomical differences to stem xylem: Mean tangential diameter of vessel lumina smaller in the branch (7-12-15/15-18-25 µm). Vessel density is higher in the stem (111-131-146/175-209-310 vessels/mm^2). Ray height is lower in branch (140-254-370/110-362-910 µm), while ray number is similar 38-41-44/36-40-45 rays/mm).

BARK ANATOMY

Branch bark: The phloem is composed by collapsed and non-collapsed sieve tubes and groups of sclereids irregular dispersed. Irregularly dispersed single sclereids in the cortex.

Stem bark: The phloem consists of alternating tangentially distributed groups of sclereids, and collapsed and non-collapsed sieve tubes. The phellem is homogeneous with dark stained substances and it consists of regularly arranged rectangular cells, indistinct in polarized light.

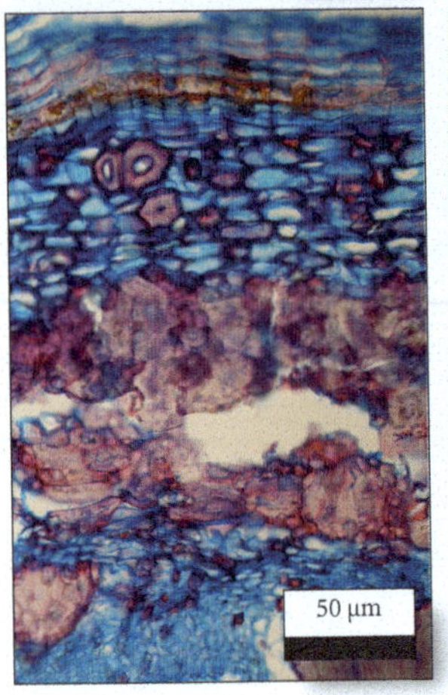

IAWA feature numbers

2 5 9 9.1 13 22 25 30 40 50 53 58 61 66 69 70 76 77 92
96.1 102.3 107 116 136 142 180 189 192 196

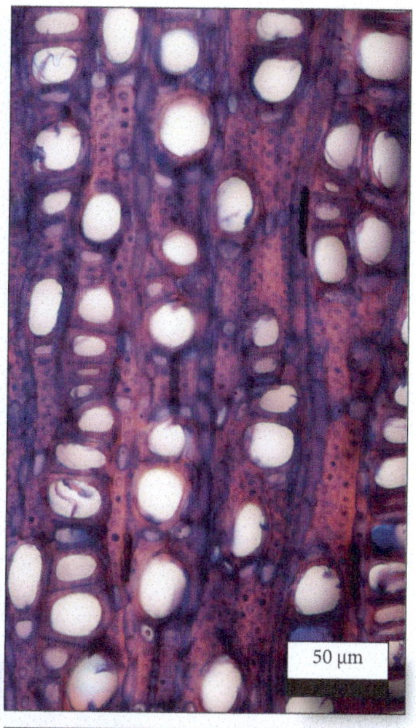

STEM XYLEM ANATOMY

Transverse section

Growth rings boundaries indistinct or absent.
Wood diffuse-porous.
Vessels solitary and in radial multiples of 2 to 4 elements.
Tangential diameter of vessels small 15-18-25 µm, 175-209-310 vessels/mm^2.
Gums and/or other deposits in heartwood vessels.
Fibers very thick-walled.
Axial parenchyma diffuse and diffuse-in-aggregates.
Rays 36-40-45/mm.

Radial section

Simple perforation plates.
Intervessel pits circular and alternate, 4-7 µm in diameter.
Vessel-ray pits with distinct borders; similar to intervessel pits in size and shape throughout the ray cell.
Vessel element length 250-353-440 µm.
Fibres with simple to minutely bordered pits (libriform fibers).
Heterogeneous rays, body ray cells procumbent with 1 to 4 rows of upright and square marginal cells.
Prismatic crystal in chambered axial parenchyma cells.

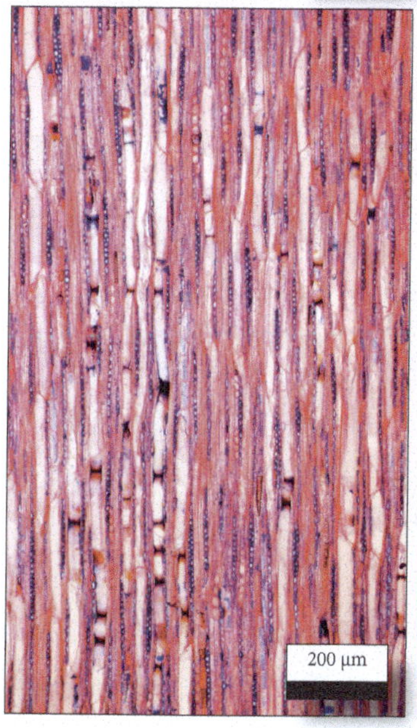

Tangential section

Rays uni- and biseriate, predominantly biseriate.
Ray height 110-362-910 µm.
Two to 4 cells per parenchyma strand.

Diospyros ferrea (Willd.) Bakh.

Synonymy: *D. buxifolia* Thouars; *D. ferrea* var. *madagascariensis* (A. DC.) Bakh.; *Ehretia ferrea* Willd.; *Maba angustifolia* Miq.; *M. buxifolia* (Rottb.) Juss.; *M. caningiana* A.DC.; *M. ebenus* Wight; *M. ferrea* (Willd.) Aubrév.; *M. guineensis* (Schumach. & Thonn.) A. DC.; *M. littorea* R. Br.; *M. madagascariensis* A. DC.; *M. smeathmanni* A. DC.

EN: Ebony FR: Bois d'ébène ML: Hazomainty

PLANT DESCRIPTION

Shrub to tree | 1-10 m high | 6-20 cm DBH
Native to Madagascar, common to central Africa and Asia.
Exploited species.

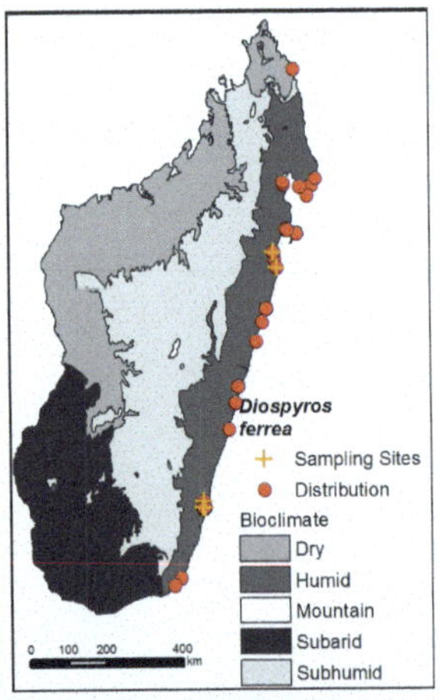

DISTRIBUTION

Elevation: 0-500 m a.s.l.
Vegetation formation: Forest
Bioclimate: Humid
Protected areas: Masoala
New protected areas: Andranotsara, Mahabo-Mananivo

Slides made from material collected in Andranotsara.

PITH AND PRIMARY XYLEM ANATOMY

Pith: Shape of pith round. Cells dimorphic. Pits in transverse cell walls. Pith visible in polarized light.

Primary xylem: Vascular bundles not distinct. Vessels solitary and in short radial multiples.

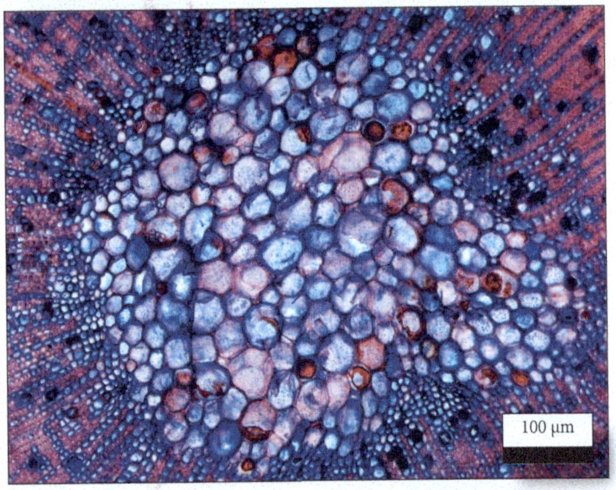

BRANCH XYLEM ANATOMY

Description: Growth rings boundaries indistinct or absent. Tangential diameter of vessels 50-53-70 µm, 47-60-72 vessels/mm². Vessel element length 200-383-610 µm. Intervessel pits small, 4-7 µm in diameter. Fibres thick- to very thick-walled. Axial parenchyma diffuse-in-aggregates. Rays exclusively uniseriate. Ray height 230-491-1200 µm. Rays 23-26-31/mm.

Anatomical differences to stem xylem: Mean tangential diameter of vessel lumina is slightly smaller in the branch (50-53-70/50-58-80 µm), while vessel density is higher in the branch (47-60-72/21-40-56 vessels/mm²). Rays are higher in the branch (230-491-1200/220-420-850 µm), while ray number is similar (23-26-31/19-25-31 rays/mm).

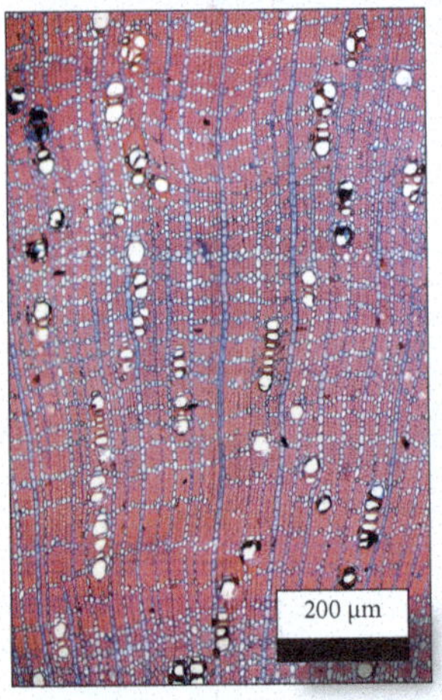

BARK ANATOMY

Branch bark: Collapsed and non-collapsed sieve tubes present in the phloem. The cortex is characterized by a band of sclereids and irregularly blue-stained parenchyma cells.

Stem bark: The phloem is characterized by collapsed and non-collapsed sieve tubes, and irregular groups of fibres. Ray and Axial parenchyma dilatation present. Prismatic crystals present in the sclereids. The cortex is characterized by bands of sclereids. The phellem cells are homogeneous and it consists of arranged rectangular cells, indistinct in polarized light

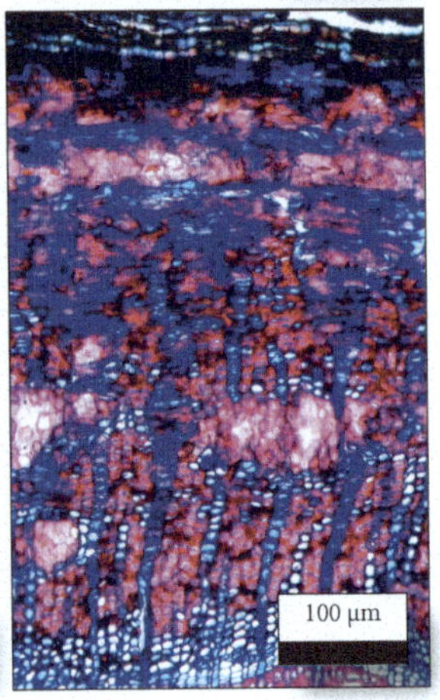

IAWA feature numbers

2 5 9.1 10 13 22 25 30 41 47 52 58 61 66 70 77 78 92 96.1 102.3 107 108 116 136 142 180 189 190 192 196

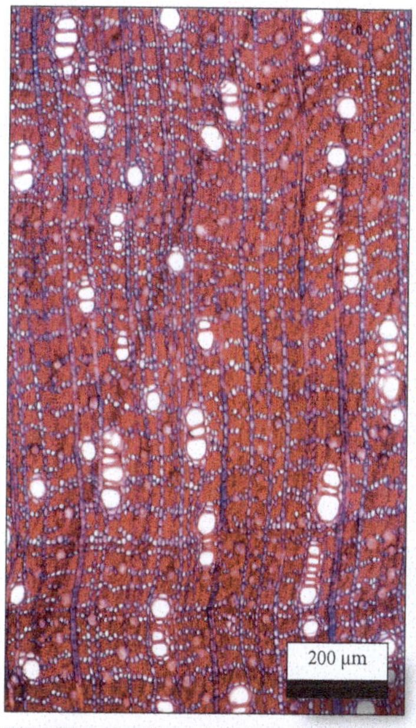

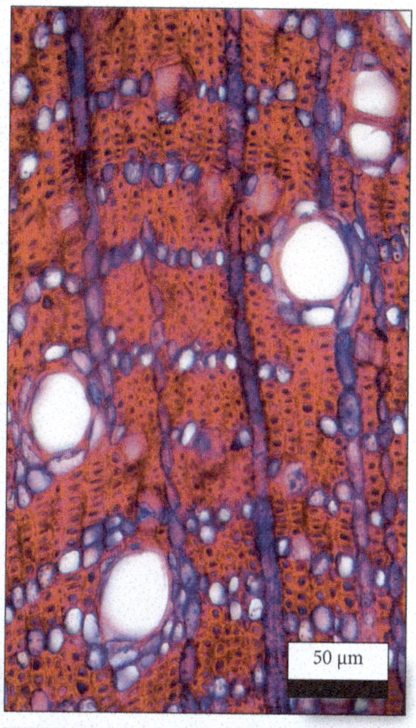

STEM XYLEM ANATOMY

Transverse section

Growth ring boundaries absent.
Wood diffuse-porous.
Vessels mostly in radial multiples of 2 to 6 elements.
Tangential diameter of vessels medium 50-58-80 μm, 21-40-56 vessels/mm^2.
Gums and/or other deposits in heartwood vessels.
Fibers thick- to very thick-walled.
Axial parenchyma diffuse-in-aggregates.
Rays 19-25-31/mm.

Radial section

Simple perforation plates.
Intervessel pits circular and alternate, 4-7 μm in diameter.
Vessel-ray pits with distinct borders; similar to intervessel pits in size and shape throughout the ray cell.
Vessel element length 120-339-450 μm.
Fibres with simple to minutely bordered pits (libriform fibers).
Heterogeneous rays, body ray cells procumbent with 2 to over 4 rows of upright and square marginal cells.
Prismatic crystal in chambered axial parenchyma cells.

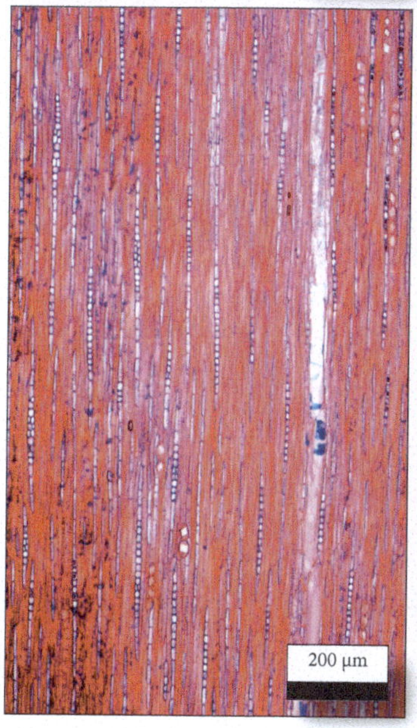

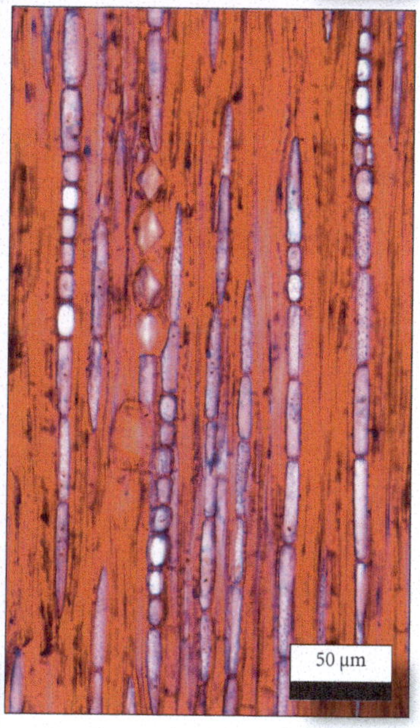

Tangential section

Rays uni- and biseriate, predominantly uniseriate.
Ray height 220-420-850 μm.
Parenchyma strand composed of 3-4 cells.

Diospyros fuscovelutina Baker.

Synonymy: *Tetraclis baroni* H. Perrier

EN: Ebony FR: Bois d'ébène ML: Hazomainty

PLANT DESCRIPTION

Shrub to tree | 2-17 m high | 4-27 cm DBH
Endemic to Madagascar. Exploited species.

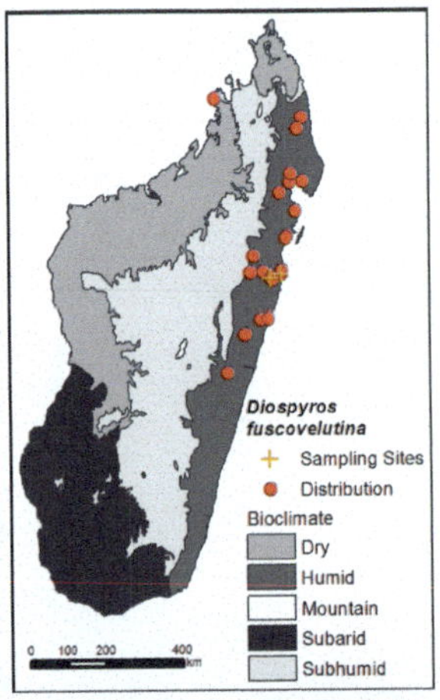

DISTRIBUTION

Elevation: 0-1500 m a.s.l.
Vegetation formation: Forest
Bioclimate: Humid and sub-humid
Protected areas: Analamazaotra-Périnet, Betampona, Mananara-Nord, Masoala, Zahamena
New protected areas: Analalava

Slides made from material collected in Betampona.

PITH AND PRIMARY XYLEM ANATOMY

Pith: Shape of pith round. Thick-walled parenchyma cells present. Cells dimorphic. Pits in transverse cell walls. Pith visible in polarized light.

Primary xylem: Vascular bundles separated. Vessels solitary and in short radial multiples.

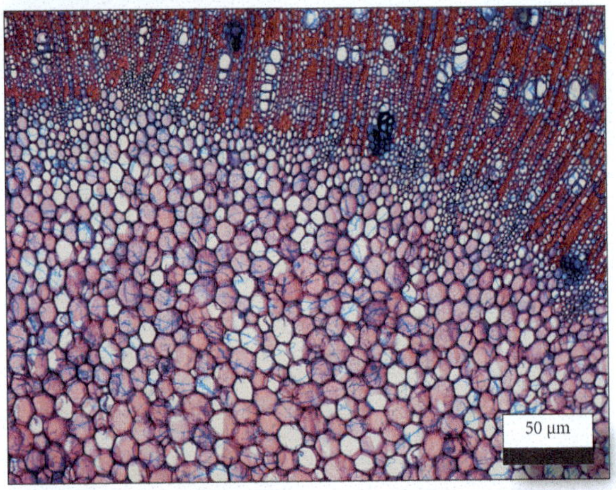

BRANCH XYLEM ANATOMY

Description: Growth rings boundaries indistinct or absent. Tangential diameter of vessels 40-49-60 µm, 25-32-45 vessels/mm². Vessel element length 210-310-450 µm. Intervessel pits small, 4-7 µm in diameter. Fibres thin- to thick-walled. Axial parenchyma diffuse-in-aggregates. Rays uni- and biseriate. Ray height 350-683-1290 µm, 31-35-40 rays/mm.

Anatomical differences to stem xylem: Mean tangential diameter of vessel lumina is smaller in the branch (40-49-60/40-52-65 µm), vessel density is slightly lower in the branch (25-32-45/19-36-38 vessels/mm²). Rays are higher in the branch (350-683-1290/270-631-1580 µm), ray number is similar (31-35-40/26-30-35 rays/mm).

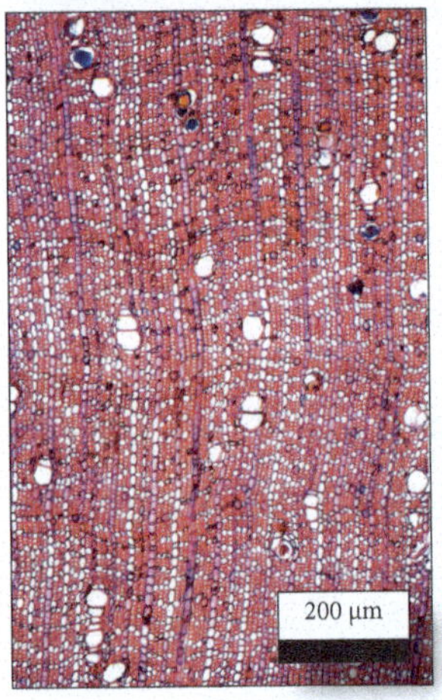

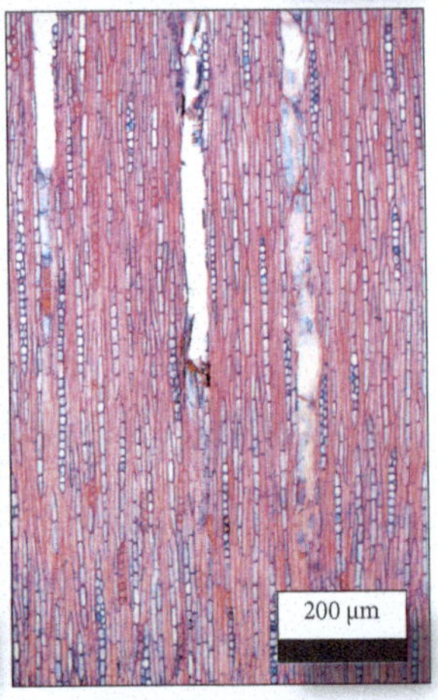

BARK ANATOMY

Branch bark: Collapsed and non-collapsed sieve tubes present in the phloem. The cortex is characterized by a band of fibers and sclereids and irregularly blue-stained parenchyma cells.

Stem bark: The phloem consists of collapsed and non-collapsed sieve tubes, and irregular groups of sclereids. Ray and Axial parenchyma dilatation present. Prismatic crystals scattered or irregularly dispersed in phloem. The cortex is characterized by a band of sclereids. The phellem cells are homogeneous and it consists of arranged rectangular cells, indistinct in polarized light.

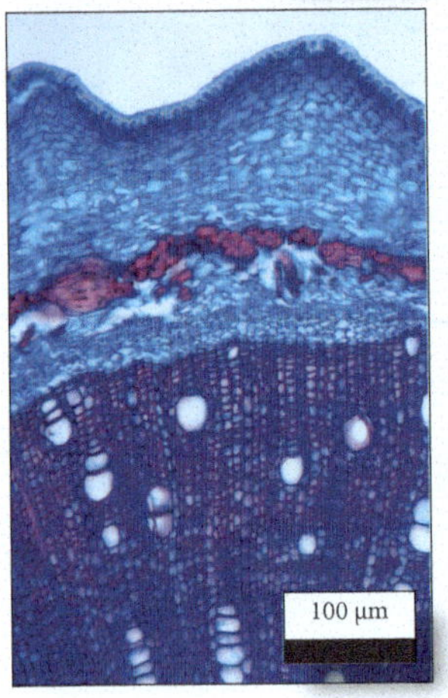

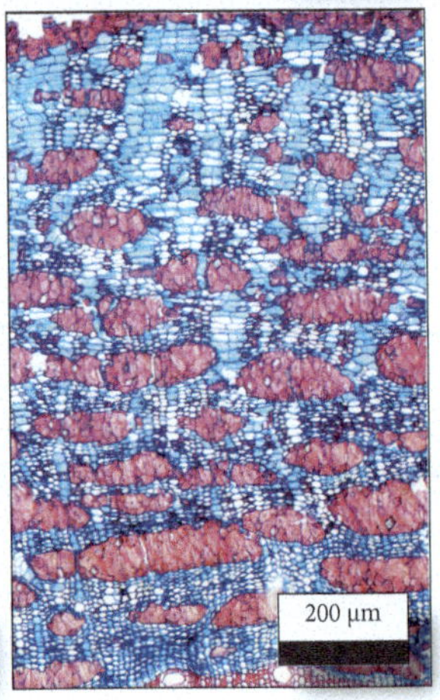

IAWA feature numbers

2 5 9.1 13 22 25 30 41 48 52 58 61 66 70 77 78 86 92 96.1 102.3 107 108 116 136 138 142 180 189 190 192 196

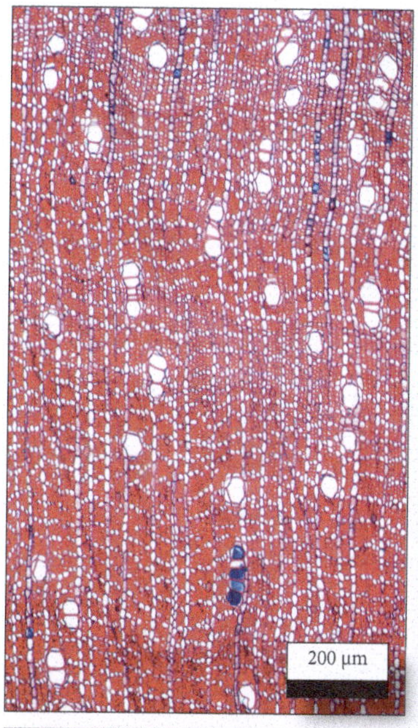

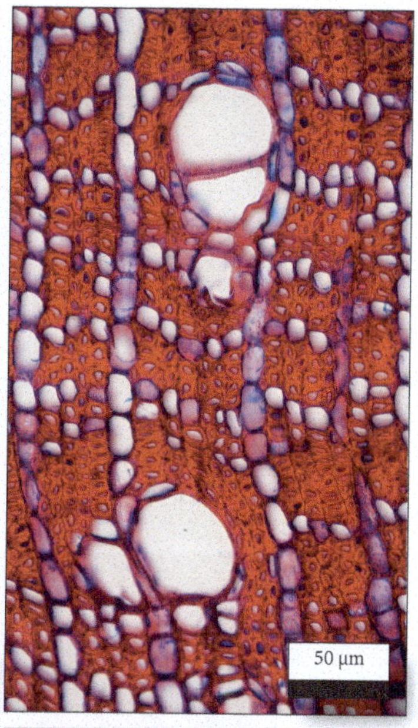

STEM XYLEM ANATOMY

Transverse section

Growth ring boundaries absent.
Wood diffuse-porous.
Vessels solitary and in short radial multiples of 2 to 4 elements.
Tangential diameter of vessels medium 40-52-65 µm, 19-36-38 vessels/mm^2.
Gums and/or other deposits in heartwood vessels.
Fibers very thick-walled.
Apotracheal parenchyma diffuse-in-aggregates and in bands up to three cells wide. Paratracheal parenchyma scanty.
Rays 26-30-35/mm.

Radial section

Simple perforation plates.
Intervessel pits circular and alternate, 4-7 µm in diameter.
Vessel-ray pits with distinct borders; similar to intervessel pits in size and shape throughout the ray cell.
Vessel element length 150-307-460 µm.
Fibres with simple to minutely bordered pits (libriform fibers).
Heterogeneous rays, body ray cells procumbent with 2 to over 4 rows and of upright and square marginal cells.
Prismatic crystal in chambered axial parenchyma cells, and in procumbent ray cells.

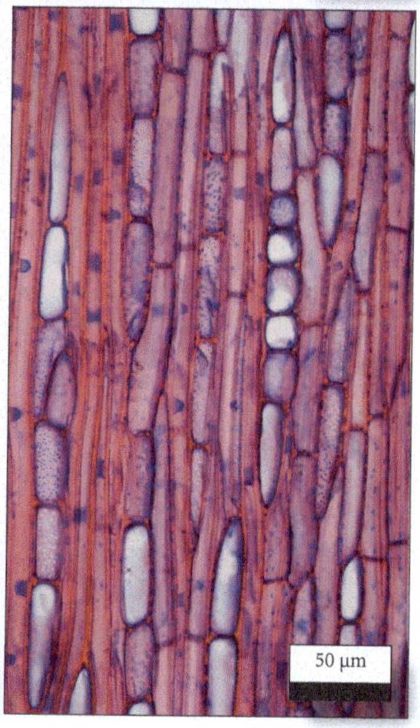

Tangential section

Rays uni- and biseriate, predominantly uniseriate.
Ray height 270-631-1580 µm.
Two to 4 cells per parenchyma strand.

Diospyros humbertiana H. Perrier

EN: Ebony FR: Bois d'ébène ML: Hazomainty

PLANT DESCRIPTION

Shrub to tree | 2-12 m high | 10-25 cm DBH
Endemic to Madagascar. Exploited species

DISTRIBUTION

Elevation: 0-1000 m a.s.l.
Vegetation formation: Forest
Bioclimate: Humid
Protected areas: Andohahela, Beza Mahafaly, Kirindy-Menabe
New protected areas: Analavelona, La Table/St Augustin, Mikea

Slides made from material collected in Kirindy-Menabe.

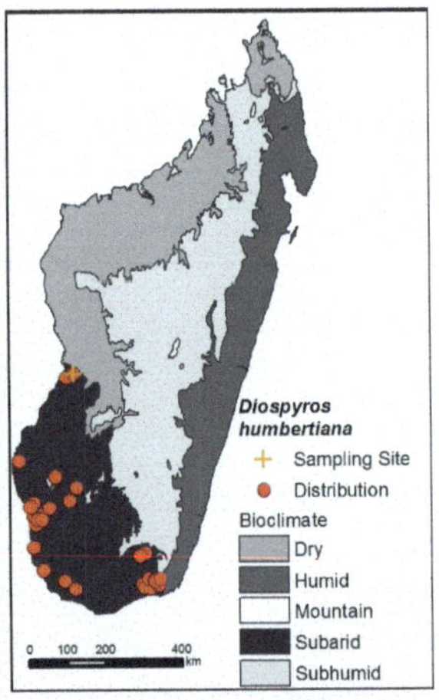

PITH AND PRIMARY XYLEM ANATOMY

Pith: Shape of pith round. Thick-walled parenchyma cells present. Cells dimorphic. Prismatic crystals present. Pits in transverse cell walls. Pith not visible in polarized light.

Primary xylem: Vascular bundles not distinct. Vessels solitary and in short radial multiples.

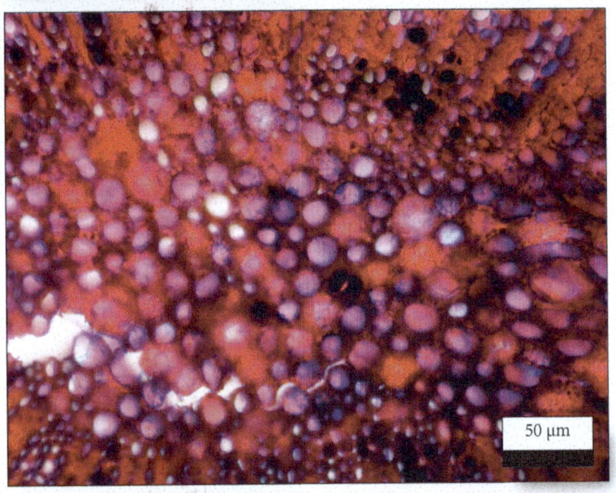

BRANCH XYLEM ANATOMY

Description: Growth rings boundaries indistinct or absent. Tangential diameter of vessels 8-11-15 µm, 110-145-190 vessels/mm^2. Vessel element length 240-294-320 µm. Intervessel pits small, 4-7 µm in diameter. Fibres thin- to thick-walled. Axial parenchyma diffuse and diffuse-in-aggregates. Rays predominantly uniseriate. Ray height 210-446-810 µm, 17-18-22 rays/mm.

Anatomical differences to stem xylem: Mean tangential diameter of vessel lumina and vessels density are slightly smaller in the branch and in the stem (8-11-15/10-18-24 µm; 110-145-190/130-160-195 vessels/mm^2). Rays are higher in the branch (210-446-810/180-422-890 µm), ray number is slightly lower in the branch (17-18-22/18-22-26 rays/mm).

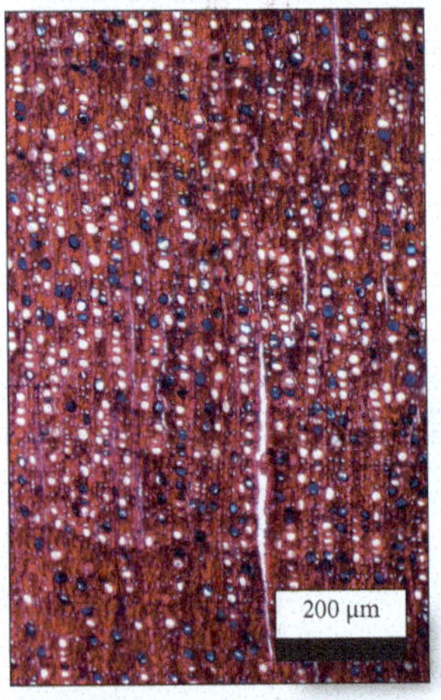

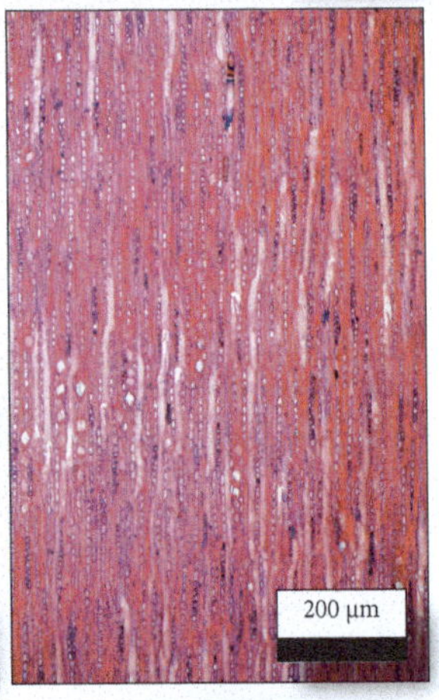

BARK ANATOMY

Branch bark: Collapsed and non-collapsed sieve tubes present in the phloem. Irregularly distributed solitary and group of sclereids present. Some rays become dilated. The cortex is characterized by a band of sclereids and irregularly blue-stained parenchyma cells.

Stem bark: The phloem consists of alternating irregular groups of sclereids, and collapsed and non-collapsed sieve tubes. Prismatic crystals present in the sclereids. The phellem is homogeneous and it consists of regularly arranged rectangular cells, indistinct in polarized light.

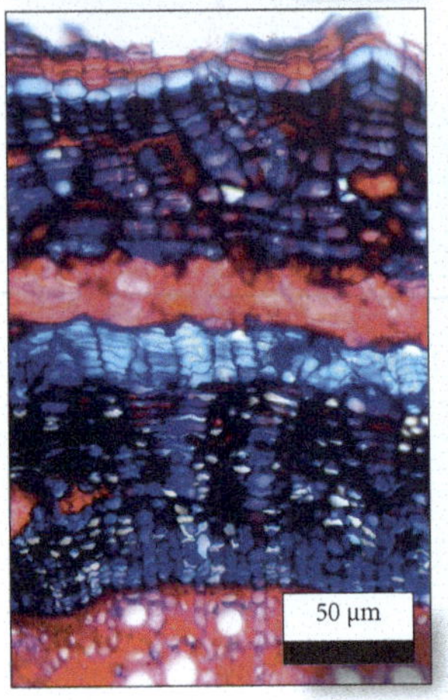

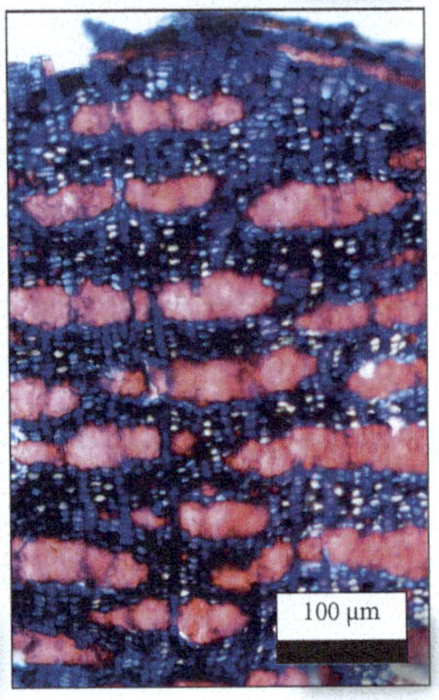

IAWA feature numbers
2 5 9 9.1 13 22 30 40 50 52 56 58 61 66 70 76 77 92 96.1
102.3 107 116 136 142 180 189 192 196

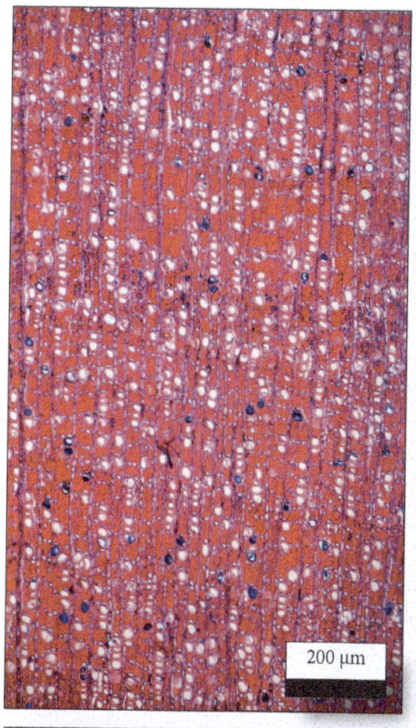

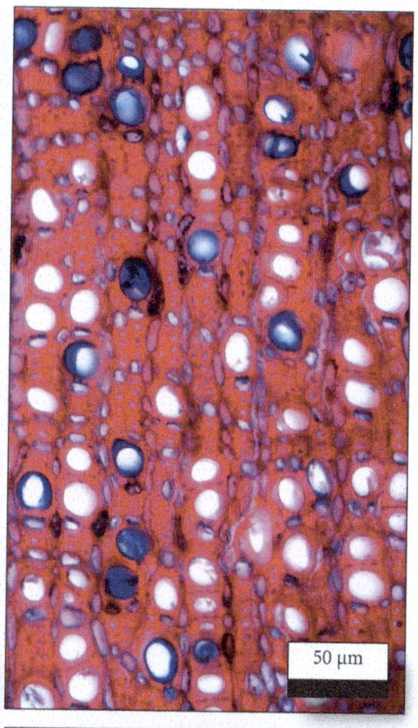

STEM XYLEM ANATOMY

Transverse section
Growth ring boundaries absent.
Wood diffuse-porous.
Vessels solitary and in short radial multiples.
Tangential diameter of vessels small 10-18-24 µm, 130-160-195 vessels/mm^2.
Tyloses common.
Gums and/or other deposits in heartwood vessels.
Fibers very thick-walled.
Axial parenchyma diffuse and diffuse-in-aggregates.
Rays 18-22-26/mm.

Radial section
Simple perforation plates.
Intervessel pits circular and alternate, 4-7 µm in diameter.
Vessel-ray pits with distinct borders; similar to intervessel pits in size and shape throughout the ray cell.
Vessel element length 210-293-320 µm.
Fibres with simple to minutely bordered pits (libriform fibers).
Heterogeneous rays, body ray cells procumbent with 1 to 4 rows of upright and square marginal cells.
Prismatic crystal frequent in chambered axial parenchyma cells.

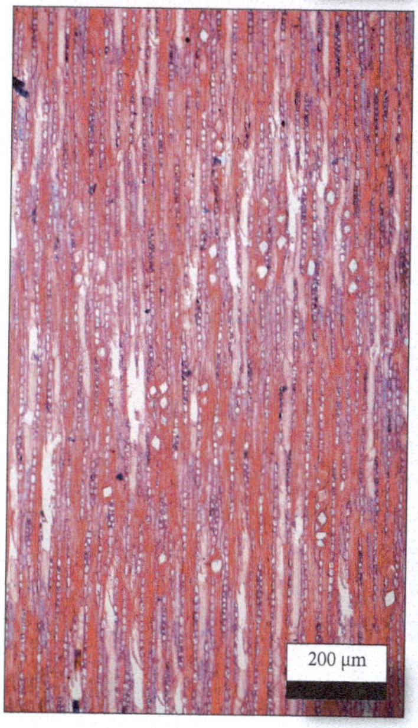

Tangential section
Rays uni- and biseriate, predominantly uniseriate.
Ray height 180-422-890 µm.
Two to 4 cells per parenchyma strand.

Diospyros latispathulata H. Perrier

EN: Ebony FR: Bois d'ébène ML: Hazomainty

PLANT DESCRIPTION

Shrub to tree | 7-25 m high | 12-30 cm DBH
Endemic to Madagascar. Exploited species.

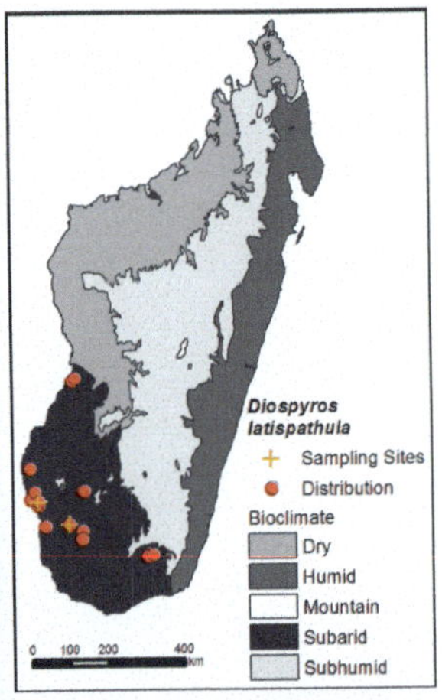

DISTRIBUTION

Elevation: 0-1500 m a.s.l.
Vegetation formation: Forest
Bioclimate: Humid and sub-humid
Protected areas: Ankarana, Lokobe, Mananara-Nord, Marojejy, Masoala, Nosy Mangabe and Zahamena
New protected areas: Ile Ste Marie and Montagne des Français

Slides made from material collected in Ankarana.

PITH AND PRIMARY XYLEM ANATOMY

Pith: Shape of pith round. Thick-walled parenchyma cells present. Cells dimorphic. Pits in transverse cell walls. Pith not visible in polarized light.

Primary xylem: Vascular bundles separated. Vessels solitary and in short radial multiples.

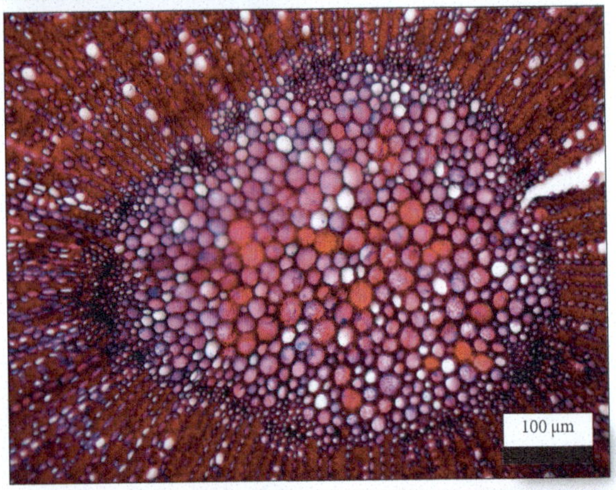

BRANCH XYLEM ANATOMY

Description: Growth rings boundaries indistinct or absent. Tangential diameter of vessels 12-17-22 µm, 111-131-146 vessels/mm^2. Vessel element length 210-286-390 µm. Intervessel pits small, 4-7 µm in diameter. Fibers thick- to very thick-walled. Axial parenchyma diffuse and diffuse-in-aggregates. Rays exclusively uniseriate, ray height 100-383-840 µm, 14-16-19 rays/mm.

Anatomical differences to stem xylem: Mean tangential diameter of vessel lumina is similar in the branch and in the stem (12-17-22/8-15-22 µm). Vessel density is slightly lower in the branch (111-131-146/125-137-148 vessels/mm^2). Ray height is higher in the branch (100-383-840/100-304-540 µm), while ray number is similar (14-16-19/13-15-18 rays/mm).

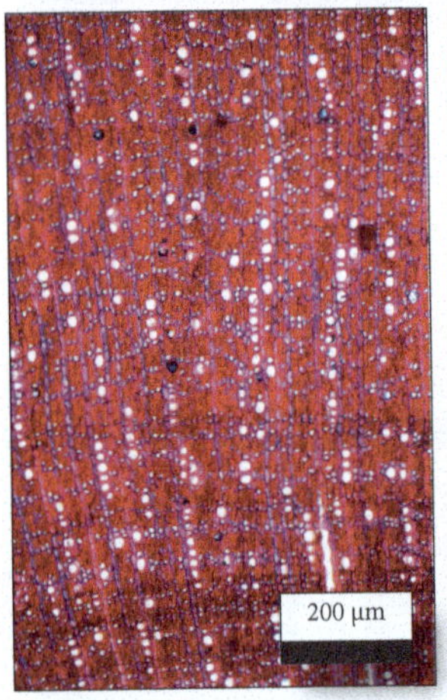

BARK ANATOMY

Branch bark: Collapsed and non-collapsed sieve tubes present in the phloem. Some rays become dilated. The cortex is characterized by a band of sclereids and irregularly blue-stained parenchyma cells

Stem bark: The phloem is characterized by collapsed and non-collapsed sieve tubes, and tangentially distributed groups of sclereids. The phellem is homogeneous with dark staining substances and it consists of regularly arranged rectangular cells, indistinct in polarized light.

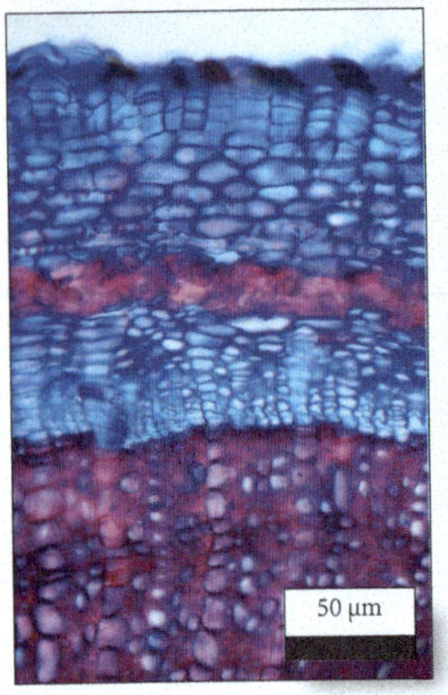

IAWA feature numbers

2 5 9.1 10 13 22 25 30 40 50 52 58 61 66 70 76 77 92 96 102.3 107 108 116 136 138 142 180 189 190 192 196

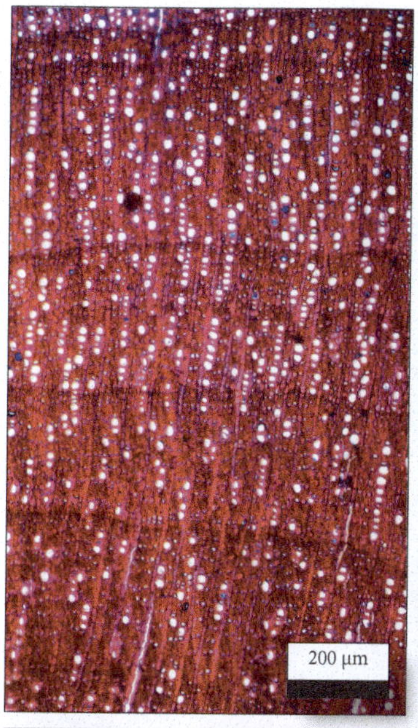

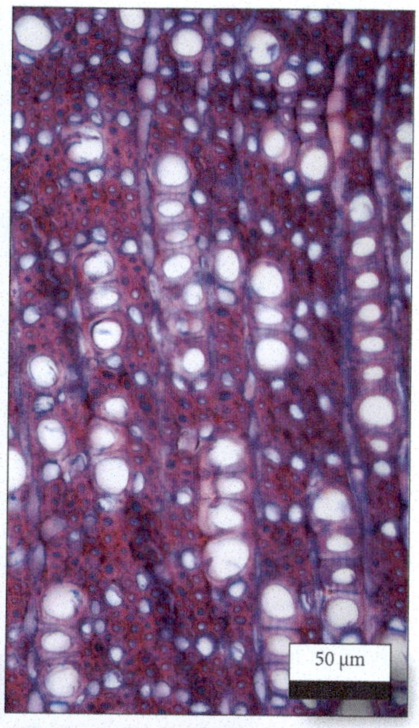

STEM XYLEM ANATOMY

Transverse section

Growth rings boundaries indistinct or absent.

Wood diffuse-porous.

Vessels solitary and in radial multiples of 2 to 8 elements.

Tangential diameter of vessels small 8-15-22 µm, 125-137-148 vessels/mm^2.

Gums and/or other deposits in heartwood vessels.

Fibers very thick-walled.

Axial parenchyma diffuse and diffuse-in-aggregates.

Rays 13-15-18/mm.

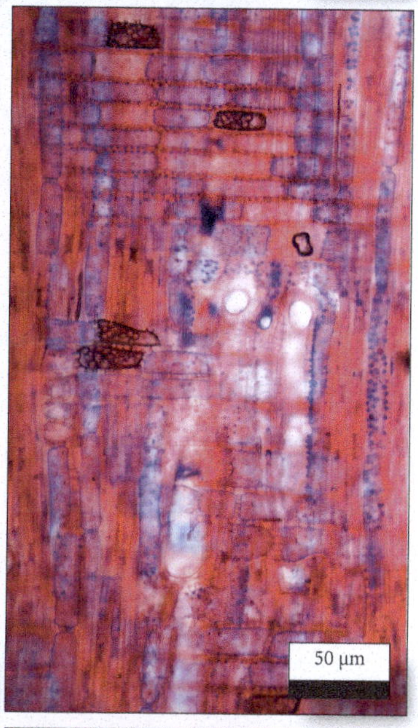

Radial section

Simple perforation plates.

Intervessel pits circular and alternate, 4-7 µm in diameter.

Vessel-ray pits with distinct borders; similar to intervessel pits in size and shape throughout the ray cell.

Vessel element length 150-251-310 µm.

Fibres with simple to minutely bordered pits (libriform fibers).

Heterogeneous rays, body ray cells procumbent with 1 to over 4 rows of upright and square marginal cells.

Prismatic crystal in chambered axial parenchyma cells, and in procumbent ray cells.

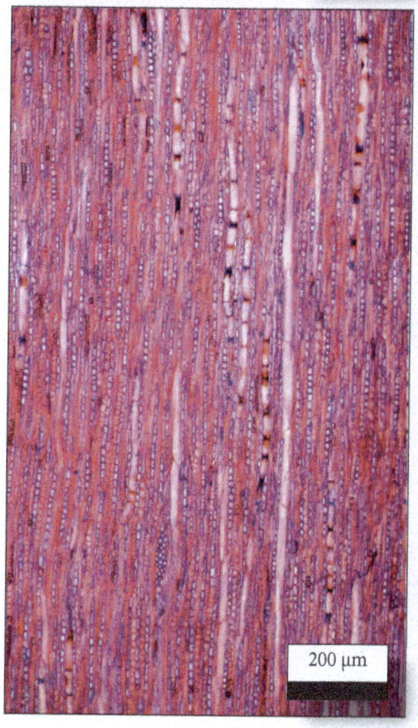

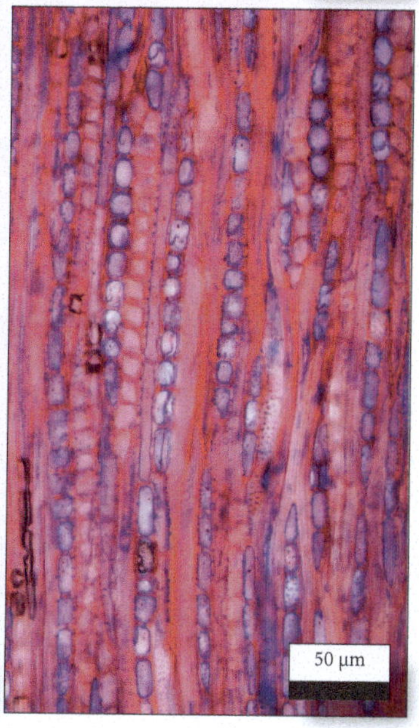

Tangential section

Rays exclusively uniseriate.

Ray height 100-304-540 µm.

Two to 4 cells per parenchyma strand.

Diospyros leucocalyx Hiern

Synonymy: *D. megasepala* Baker

EN: Ebony FR: Bois d'ébène ML: Hazomainty

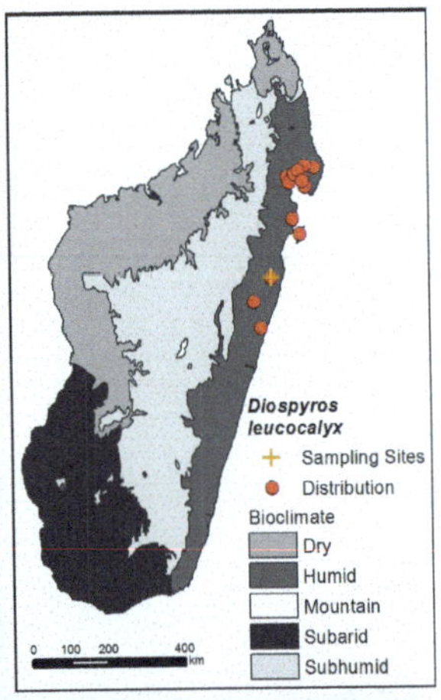

PLANT DESCRIPTION

Shrub to tree | 3-7 m high | 2-7 cm DBH
Endemic to Madagascar. Non-exploited species.

DISTRIBUTION

Elevation: 0-500 m a.s.l.
Vegetation formation: Forest
Bioclimate: Humid
Protected areas: Betampona, Mananara-Nord, Masoala, Nosy Mangabe
New protected areas: Ankerana and Ile Ste Marie

Slides made from material collected in Betampona.

PITH AND PRIMARY XYLEM ANATOMY

Pith: Shape of pith round. Cells dimorphic. Prismatic crystals present. Pits in transverse cell walls. Pith not visible in polarized light.

Primary xylem: Vascular bundles separated. Vessels solitary and in short radial multiples.

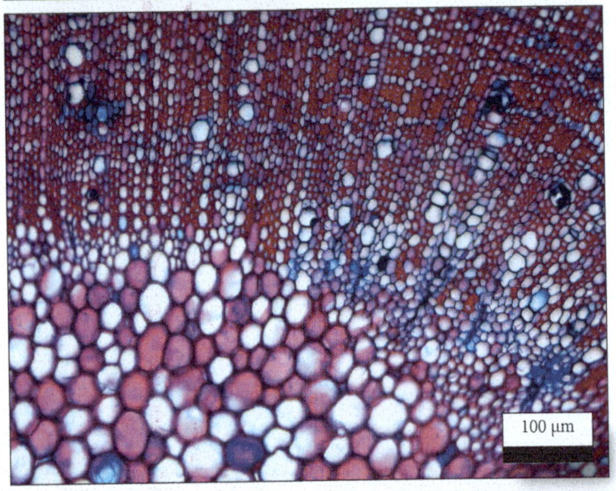

BRANCH XYLEM ANATOMY

Description: Growth rings boundaries indistinct or absent. Tangential diameter of vessels 43-51-64 µm, 25-29-32 vessels/mm^2. Vessel element length 320-442-580 µm. Intervessel pits small, 4-7 µm in diameter. Fibres thin- to thick-walled. Axial parenchyma diffuse-in-aggregates and in narrow lines up to 3 cells. Rays exclusively uniseriate, ray height 200-459-1020 µm. Rays 30-35-42/mm.

Anatomical differences to stem xylem: Mean tangential diameter of vessel lumina is slightly smaller in the branch (43-51-64/50-62-80 µm). Vessel density is higher in the branch (25-29-32/15-21-31 vessels/mm^2). Rays are lower in the branch (200-459-1020/120-690-1050 µm). Ray number is higher in the branch (30-35-42/17-20-22 rays/mm).

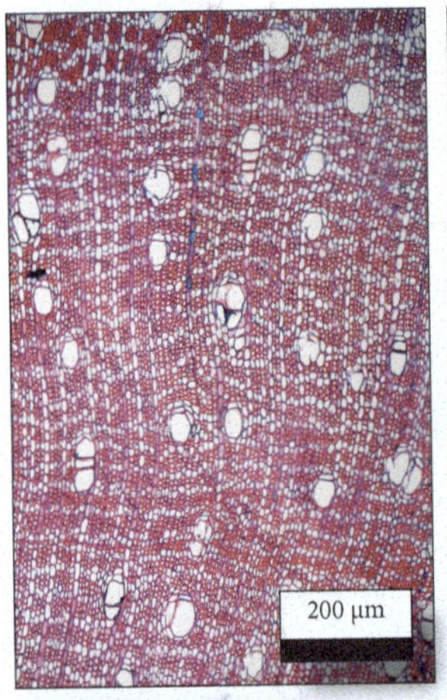

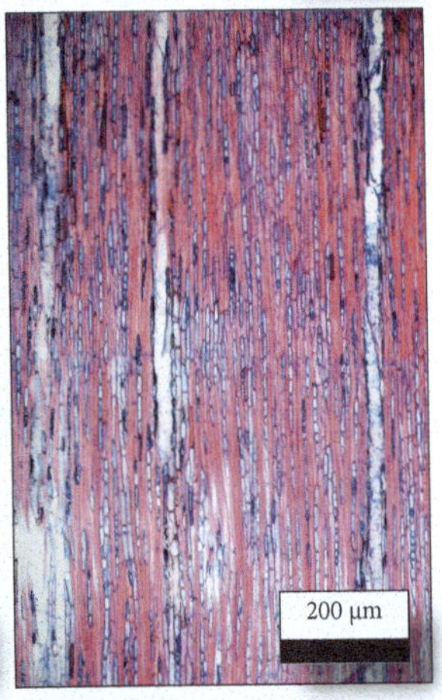

BARK ANATOMY

Branch bark: Collapsed and non-collapsed sieve tubes present in the phloem. The cortex is characterized by a band of sclereids and irregularly blue-stained parenchyma cells.

Stem bark: The phloem is characterized by collapsed and non-collapsed sieve tubes and irregularly distributed groups of sclereids. Some rays became dilated. The cortex is characterized by a band of sclereids and irregularly blue-stained parenchyma cells. The phellem is homogeneous with dark staining substances and it consists of regularly arranged rectangular cells, indistinct in polarized light.

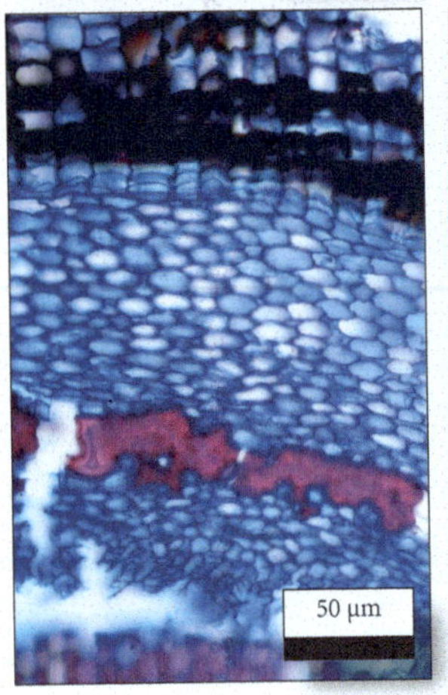

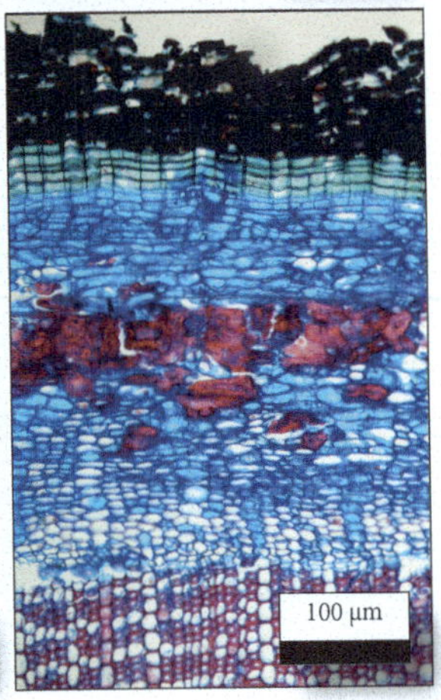

IAWA feature numbers

2 5 9.1 13 22 25 30 41 48 53 58 61 66 69 77 78 92 96
102.4 109 116 136 138 142 180 189 190 192 196

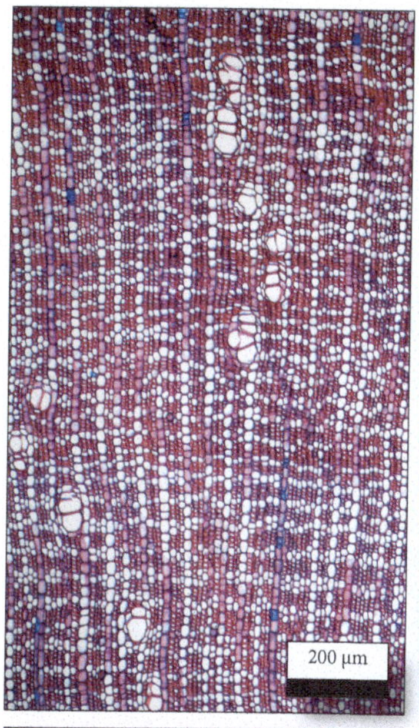

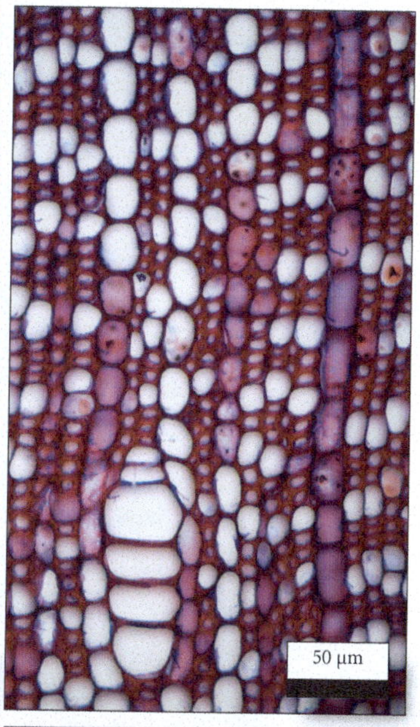

STEM XYLEM ANATOMY

Transverse section

Growth rings boundaries indistinct or absent.
Wood diffuse-porous.
Vessels solitary and in short radial multiples of 2 to 4 elements.
Tangential diameter of vessels medium 50-62-80 µm, 15-21-31 vessels/mm^2.
Gums and/or other deposits in heartwood vessels.
Fibers thin- to thick-walled.
Apotracheal parenchyma diffuse-in-aggregates, paratracheal parenchyma scanty.
Rays 17-20-22/mm.

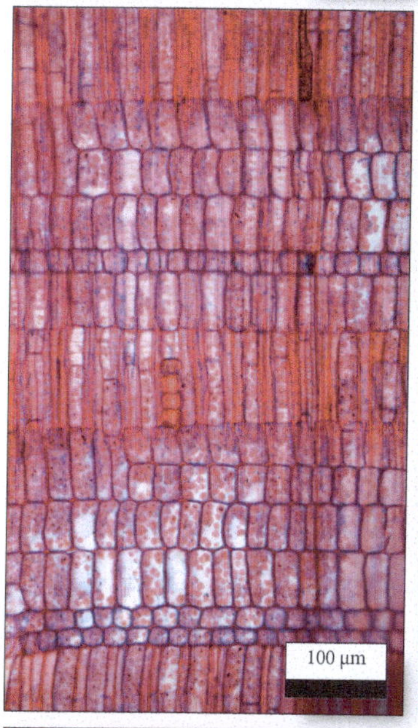

Radial section

Simple perforation plates.
Intervessel pits circular and alternate, 4-7 µm in diameter.
Vessel-ray pits with distinct borders; similar to intervessel pits in size and shape throughout the ray cell.
Vessel element length 320-450-650 µm.
Fibres with simple to minutely bordered pits (libriform fibers).
Heterogeneous rays, with procumbent, square and upright cells mixed throughout the ray.
Prismatic crystal in chambered axial parenchyma cells.

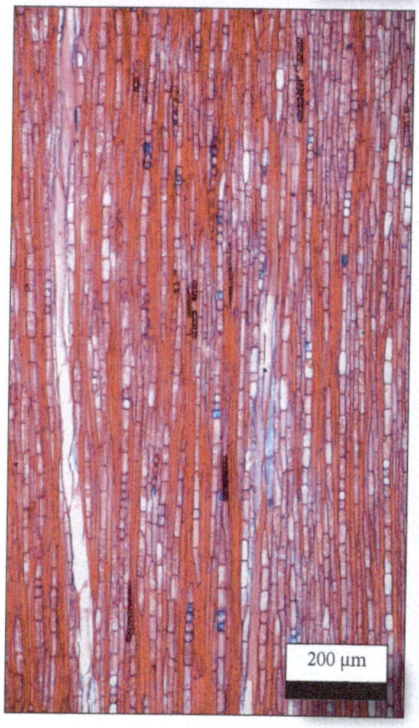

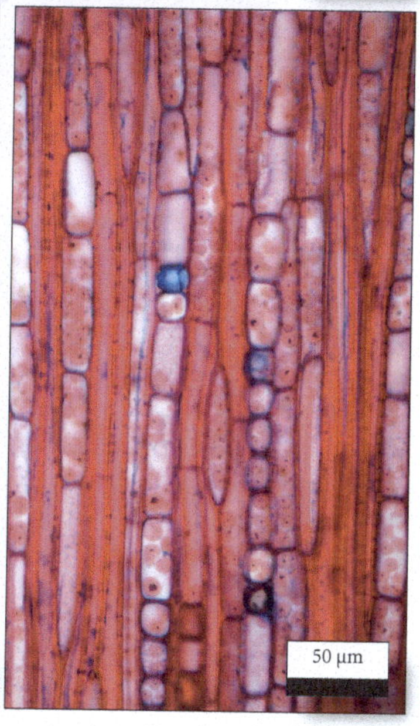

Tangential section

Rays exclusively uniseriate.
Ray height 120-690-1050 µm.
Two to 4 cells per parenchyma strand.

Diospyros maculata ined.

EN: Ebony FR: Bois d'ébène ML: Hazomainty

PLANT DESCRIPTION

Tree | 5-25 m high | 25 cm DBH
Endemic to Madagascar. Non-exploited species.

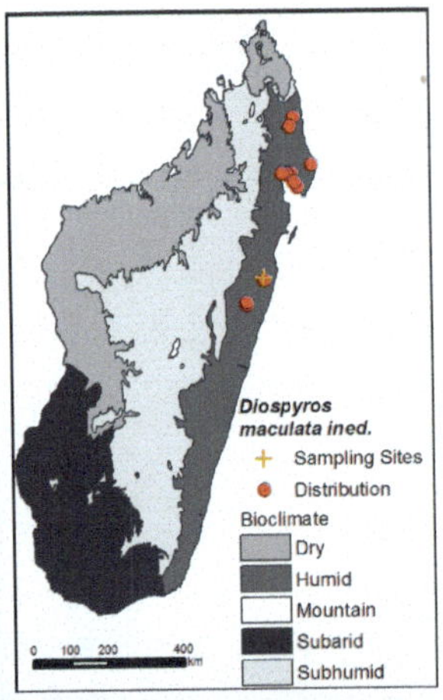

DISTRIBUTION

Elevation: 0-1000 m a.s.l.
Vegetation formation: Forest
Bioclimate: Humid
Protected areas: Nosy Mangabe, Betampona and Masoala
New protected areas: Makirovana, Sahafary and Ankerana

Slides made from material collected in Makirovana.

PITH AND PRIMARY XYLEM ANATOMY

Pith: Shape of pith oval. Thick-walled parenchyma cells present. Cells dimorphic. Prismatic crystals present. Pits in transverse cell walls. Pith not visible in polarized light.

Primary xylem: Vascular bundles not distinct. Vessels solitary and in short radial multiples.

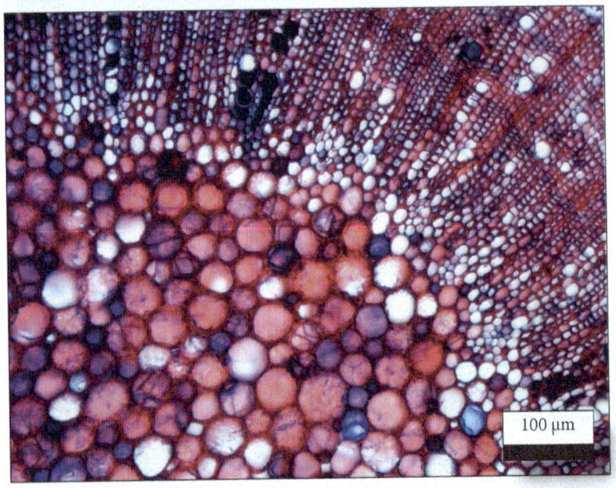

BRANCH XYLEM ANATOMY

Description: Growth rings boundaries indistinct or absent. Tangential diameter of vessels 40-63-90 µm, 39-52-65 vessels/mm^2. Vessel element length 220-291-380 µm. Intervessel pits small, 4-7 µm in diameter. Fibres thin- to thick-walled. Axial parenchyma diffuse-in-aggregates, and scanty paratracheal. Rays predominantly uniseriate. Ray height 190-527-1040 µm. Rays 35-39-41/mm.

Anatomical differences to stem xylem: Mean tangential diameter of vessel lumina is smaller in the branch (40-63-90/80-96-120 µm). Vessel density is higher in the branch (39-52-65/8-12-18 vessels/mm^2). Rays are higher in the branch (190-527-1040/160-440-760 µm), while ray number is higher in the branch (35-39-41/20-21-23 rays/mm).

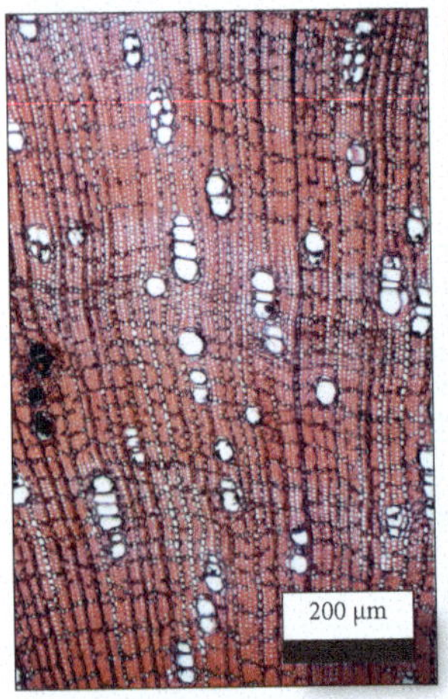

BARK ANATOMY

Branch bark: Collapsed and non-collapsed sieve tubes, and irregularly distributed groups of sclereids present in the phloem. The cortex is characterized by a band of sclereids and irregular blue and red stained parenchyma cells.

Stem bark: The phloem consists of collapsed and non-collapsed sieve tubes and irregular groups of sclereids. Some rays become dilated. The cortex is characterized by bands of sclereids. The phellem is homogeneous with dark stained substances and it consists of regularly arranged rectangular cells, indistinct in polarized light.

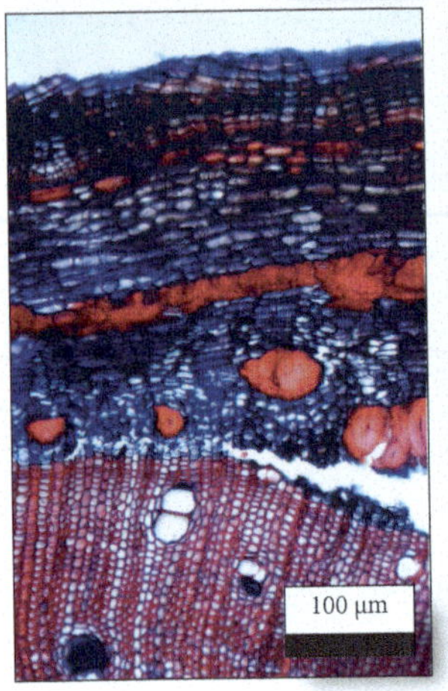

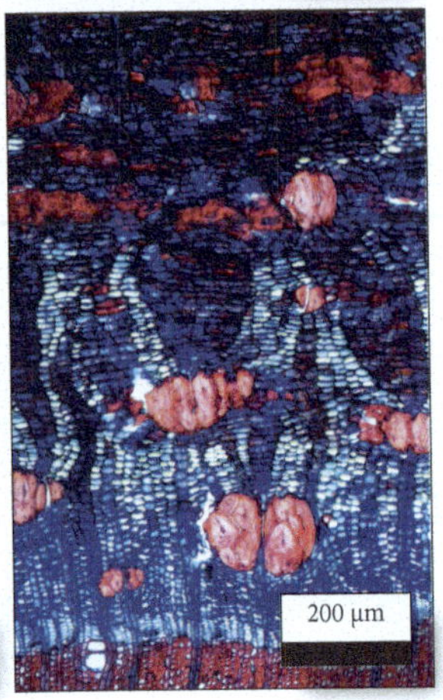

IAWA feature numbers
2 5 9.1 13 22 25 30 41 47 52 58 61 66 69 76 78 92 93 96.1 100 102.3 107 116 136 142 180 189 192 196

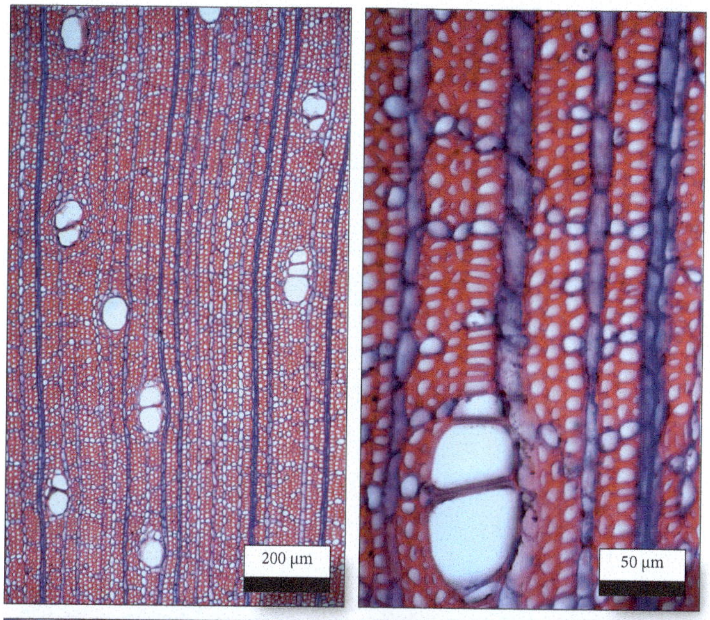

STEM XYLEM ANATOMY

Transverse section
Growth ring boundaries absent.
Wood diffuse-porous.
Vessels solitary and in short radial multiples of 2 to 4 elements.
Tangential diameter of vessels medium 80-96-120 µm, 8-12-18 vessels/mm^2.
Gums and/or other deposits in heartwood vessels.
Fibers thin- to thick-walled.
Apotracheal parenchyma diffuse, diffuse-in-aggregates, paratracheal parenchyma scanty.
Rays 20-21-23/mm.

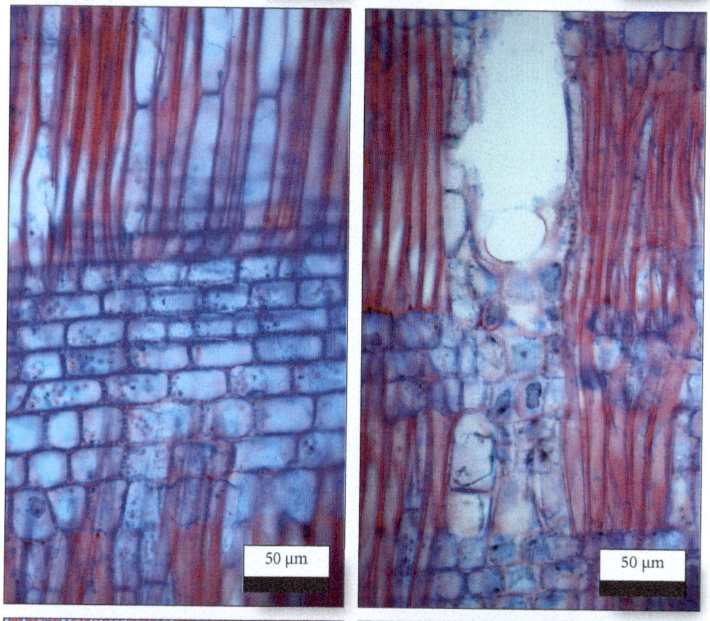

Radial section
Simple perforation plates.
Intervessel pits circular and alternate, 4-7 µm in diameter.
Vessel-ray pits with distinct borders; similar to intervessel pits in size and shape throughout the ray cell.
Vessel element length 210-307-500 µm.
Fibres with simple to minutely bordered pits (libriform fibers).
Heterogeneous rays, body ray cells procumbent with 1 to 4 rows of upright and square marginal cells.
Prismatic crystal in chambered axial parenchyma cells.

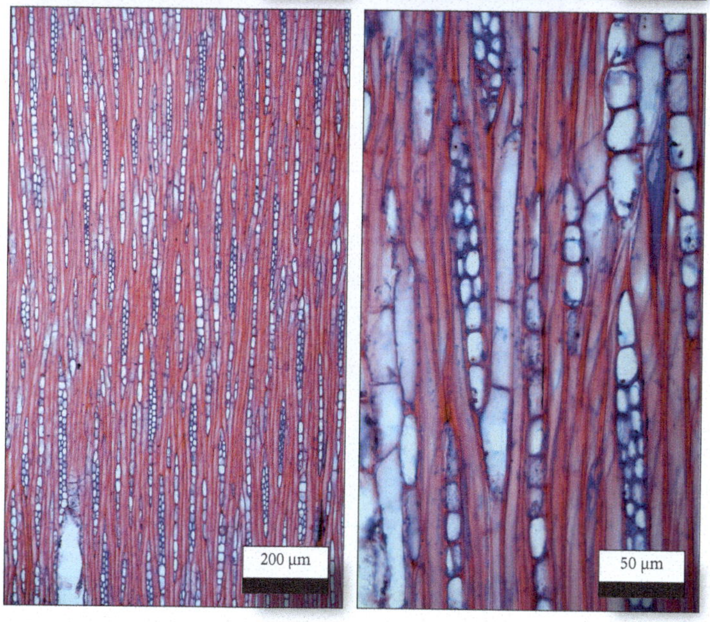

Tangential section
Rays uni- and biseriate, predominantly biseriate.
Biseriate portion as wide as uniseriate portion.
Ray height 160-440-760 µm.
Three to 8 cells per parenchyma strand.

Diospyros mangabensis Aug. DC.

EN: Ebony FR: Bois d'ébène ML: Hazomainty

PLANT DESCRIPTION

Tree | 4-18 m high | 12-60 cm DBH
Endemic to Madagascar. Exploited species.

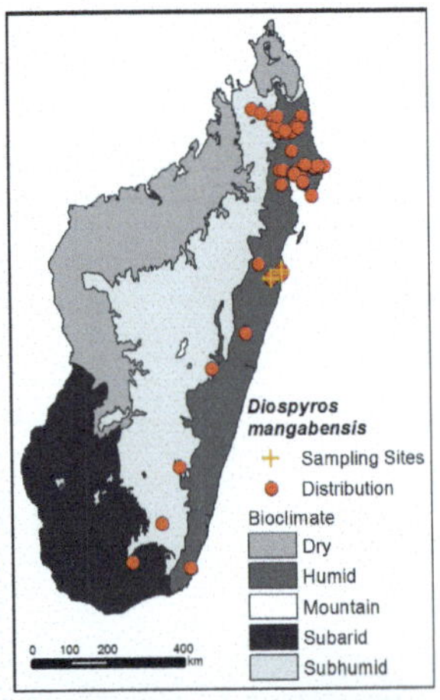

DISTRIBUTION

Elevation: 0-2000 m a.s.l.
Vegetation formation: Forest
Bioclimate: Humid and sub-humid
Protected areas: Andohahela, Andringitra, Masoala, Anjanaharibe-Sud, Betampona, Kalambatritra, Marojejy, Andasibe, Nosy Mangabe, Tsaratanana and Zahamena
New protected areas: Analalava

Slides made from material collected in Andasibe.

PITH AND PRIMARY XYLEM ANATOMY

Pith: Shape of pith oval. Thick-walled parenchyma cells present. Cells dimorphic. Pits in transverse cell walls. Pith visible in polarized light.

Primary xylem: Vascular bundles separated. Vessels solitary and in short radial multiples.

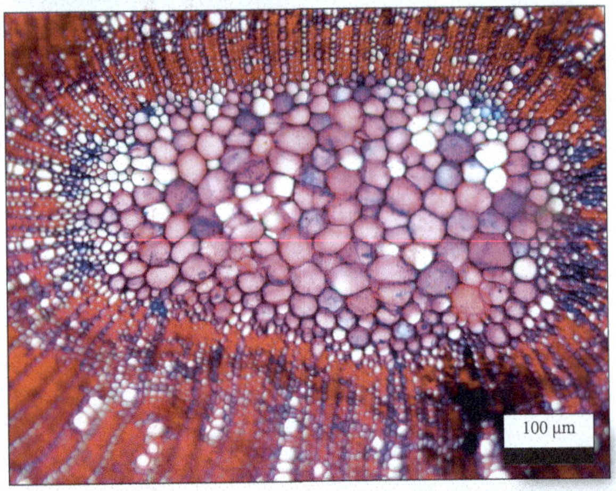

BRANCH XYLEM ANATOMY

Description: Growth rings boundaries indistinct or absent. Tangential diameter of vessels 50-64-76 µm, 19-29-49 vessels/mm². Vessel element length 290-377-540 µm. Intervessel pits small, 4-7 µm in diameter. Fibres thin- to thick-walled. Axial parenchyma diffuse-in-aggregates. Rays predominantly uniseriate. Ray height 110-367-670 µm. Rays 32-36-39/mm.

Anatomical differences to stem xylem: Mean tangential diameter of vessel lumina is smaller in the branch (50-64-76/62-78-95 µm) while vessel density is higher in the branch (19-29-49/15-21-23 vessels/mm²). Rays height and number are lower in the branch (110-367-670/160-467-980 µm; 32-36-39/33-38-42 rays/mm).

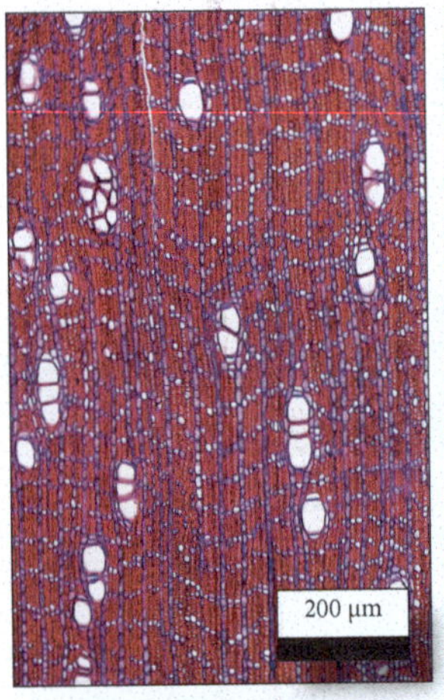

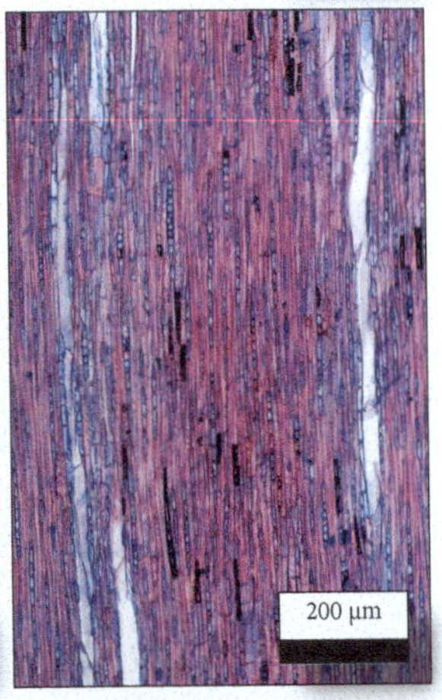

BARK ANATOMY

Branch bark: Collapsed and non-collapsed sieve tubes present in the phloem. The cortex is characterized by a band of sclereids and irregularly blue-stained parenchyma cells.

Stem bark: The phloem consists of alternating groups of tangentially distributed sclereids, and collapsed and non-collapsed sieve tubes. Ray and Axial parenchyma dilatation present. Prismatic crystals scattered or irregularly dispersed in phloem. The cortex is characterized by a band of sclereids. The phellem cells are homogeneous with dark staining substances and it consists of arranged rectangular cells, indistinct in polarized light.

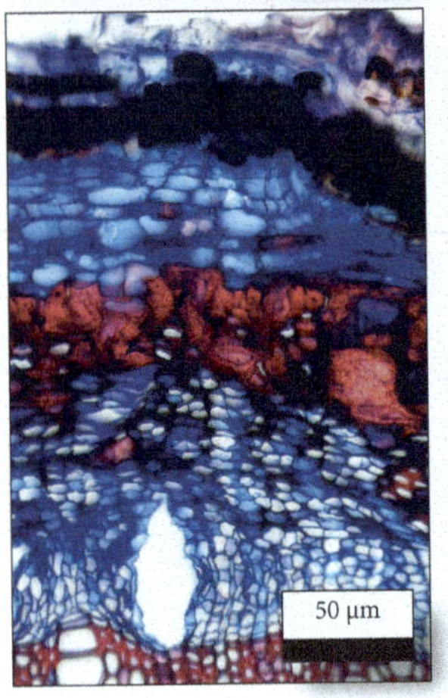

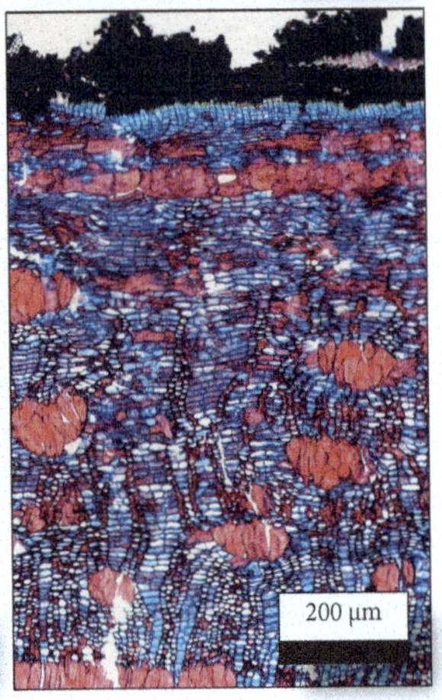

IAWA feature numbers
2 5 9.1 10 13 22 25 30 41 48 53 58 61 66 69 70 77 78 92
96.1 102.3 107 116 136 142 180 189 192 196

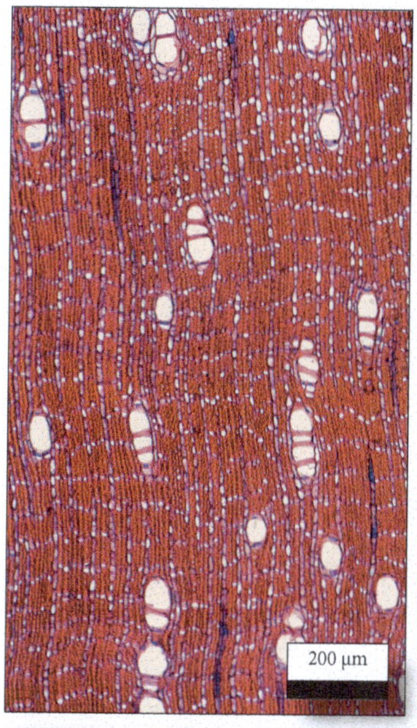

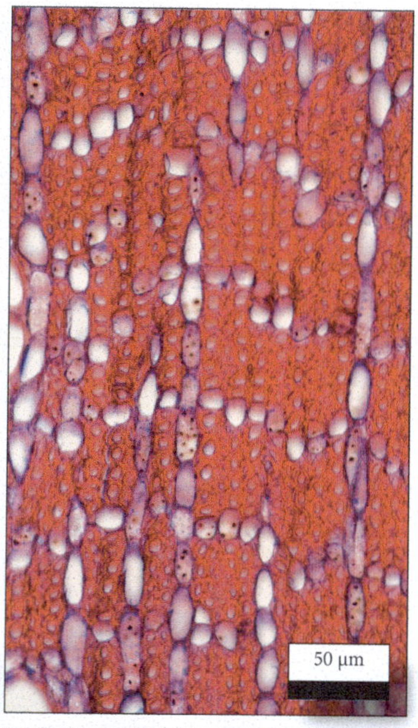

STEM XYLEM ANATOMY

Transverse section

Growth rings boundaries indistinct or absent.
Wood diffuse-porous.
Vessels solitary and in radial multiples of 2 to 6 elements.
Tangential diameter of vessels medium 62-78-95 µm, 15-21-23 vessels/mm^2.
Gums and/or other deposits in heartwood vessels.
Fibers thick- to very thick-walled.
Apotracheal parenchyma diffuse-in-aggregates, paratracheal parenchyma scanty.
Rays 33-38-42/mm.

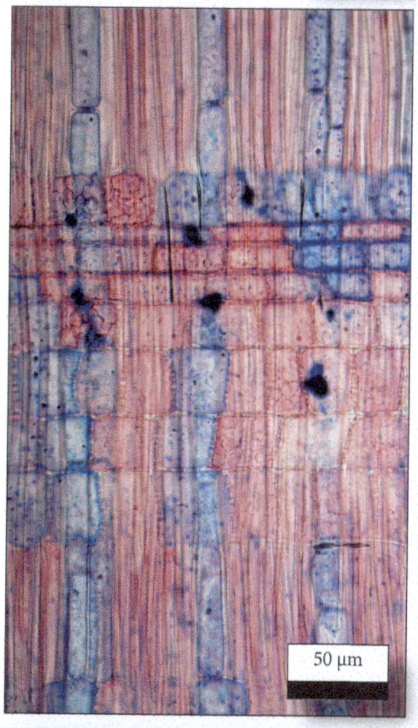

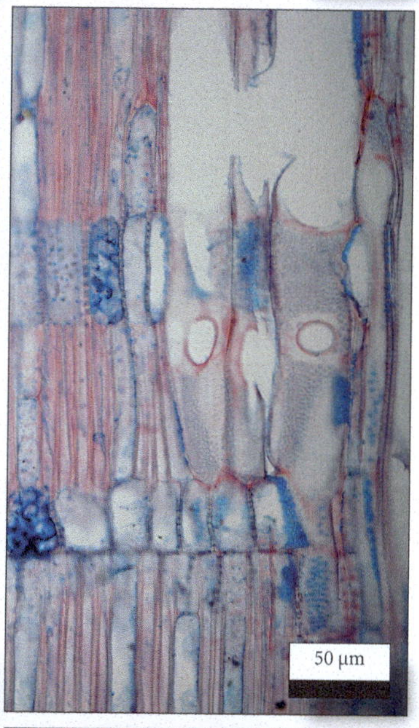

Radial section

Simple perforation plates.
Intervessel pits circular and alternate, 4-7 µm in diameter.
Vessel-ray pits with distinct borders; similar to intervessel pits in size and shape throughout the ray cell.
Vessel element length 190-349-430 µm.
Fibres with simple to minutely bordered pits (libriform fibers).
Heterogeneous rays, body ray cells procumbent with 2-4 rows of upright and square marginal cells.
Prismatic crystal in chambered axial parenchyma cells.

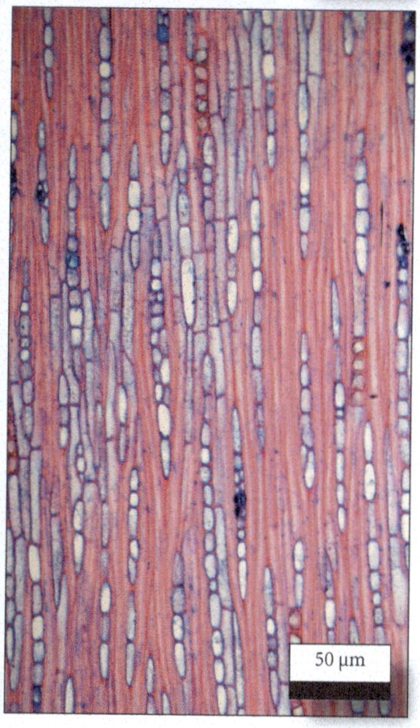

Tangential section

Rays uni- and biseriate, predominantly uniseriate.
Ray height 160-467-980 µm.
Three to 4 cells per parenchyma strand.

Diospyros maxima ined.

EN: Ebony FR: Bois d'ébène ML: Hazomainty

PLANT DESCRIPTION

Shrub to tree | 1-10 m | 6-20 cm DBH
Endemic to Madagascar. Exploited species.

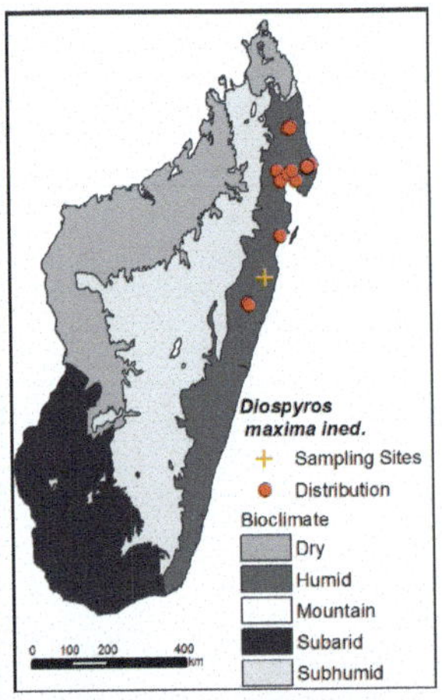

DISTRIBUTION

Elevation: 0-1000 m a.s.l.
Vegetation formation: Forest
Bioclimate: Humid
Protected areas: Masoala, Betampona and Marojejy
New protected areas: Ankerana

Slides made from material collected in Betampona.

PITH AND PRIMARY XYLEM ANATOMY

Pith: Shape of pith oval. Cells dimorphic. Pits in transverse cell walls. Pith visible in polarized light.

Primary xylem: Vascular bundles separated. Vessels solitary and in long radial multiples.

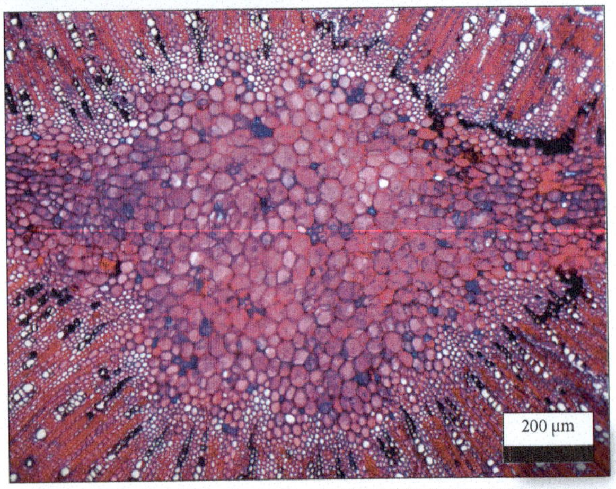

BRANCH XYLEM ANATOMY

Description: Growth rings boundaries indistinct or absent. Tangential diameter of vessels 40-63-81 µm, 23-39-52 vessels/mm². Vessel element length 221-262-342 µm. Intervessel pits small, 4-7 µm in diameter. Fibers thin- to thick-walled. Apotracheal parenchyma diffuse-in-aggregates, and paratracheal scanty paratracheal. Rays exclusively uniseriate. Ray height 170-636-1200 µm, 25-32-39 rays/mm.

Anatomical differences to stem xylem: Mean tangential diameter of vessel lumina is slightly smaller in the branch (40-63-81/60-74-90 µm), while vessel density is slightly higher in the branch (23-39-52/30-36-43 vessels/mm²). Ray height and number are higher in the branch (170-636-1200/260-505-920 µm; 25-32-39/20-25-28 rays/mm).

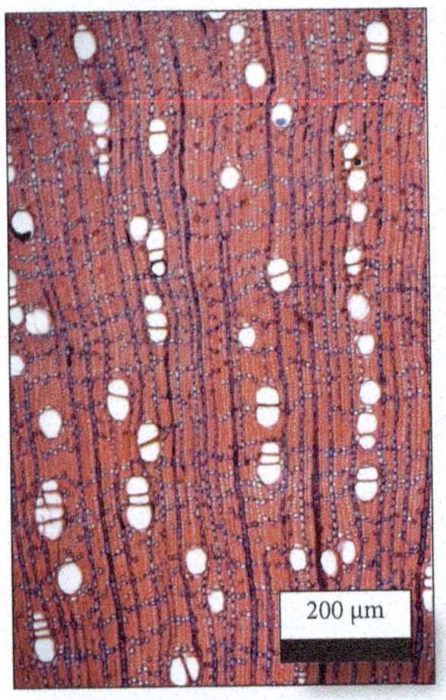

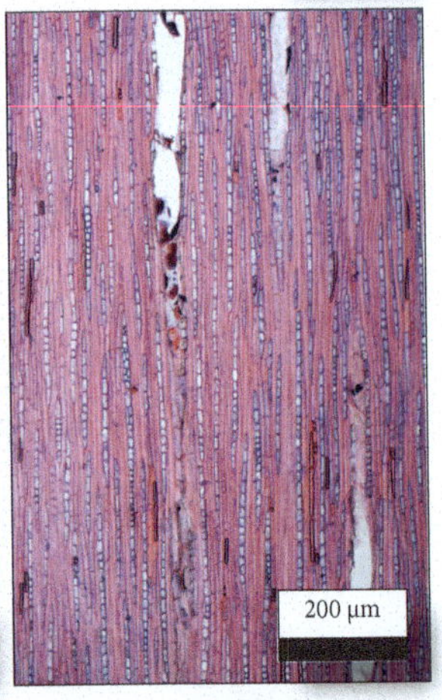

BARK ANATOMY

Branch bark: Collapsed and non-collapsed sieve tubes present in the phloem. The cortex is characterized by a band of sclereids and irregularly blue-stained parenchyma cells.

Stem bark: The phloem is characterized by collapsed and non-collapsed sieve tubes, and irregularly distributed groups of sclereids. Ray dilatation present. A band of sclereids in the cortex. The phellem is homogeneous with dark stained substances and it consists of regularly arranged rectangular cells, indistinct in polarized light.

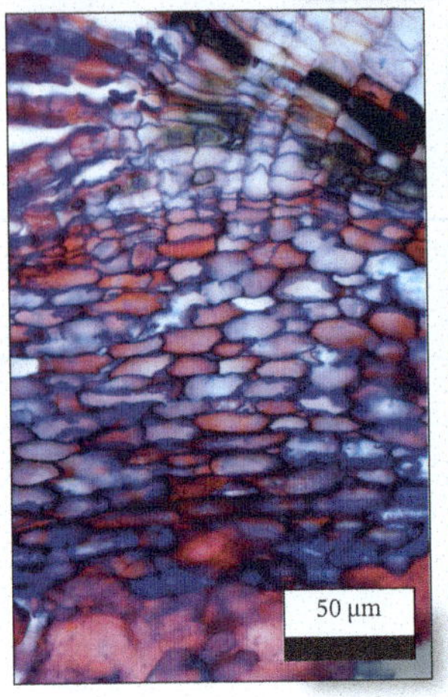

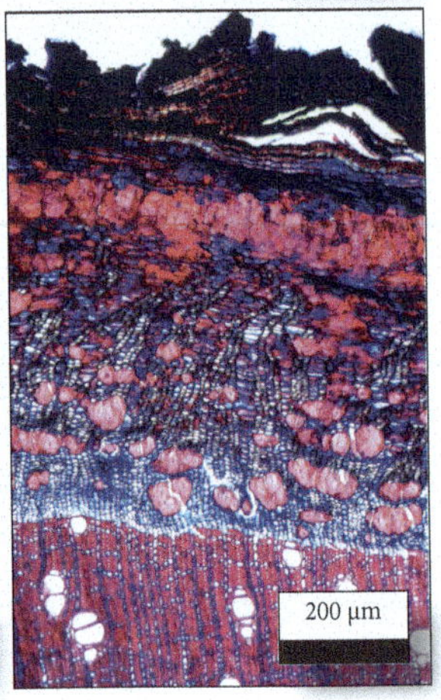

IAWA feature numbers
2 5 9.1 10 13 22 25 30 41 48 52 58 61 66 69 77 78 92
93 96 102.4 107 108 116 136 138 142 180 189 192 196

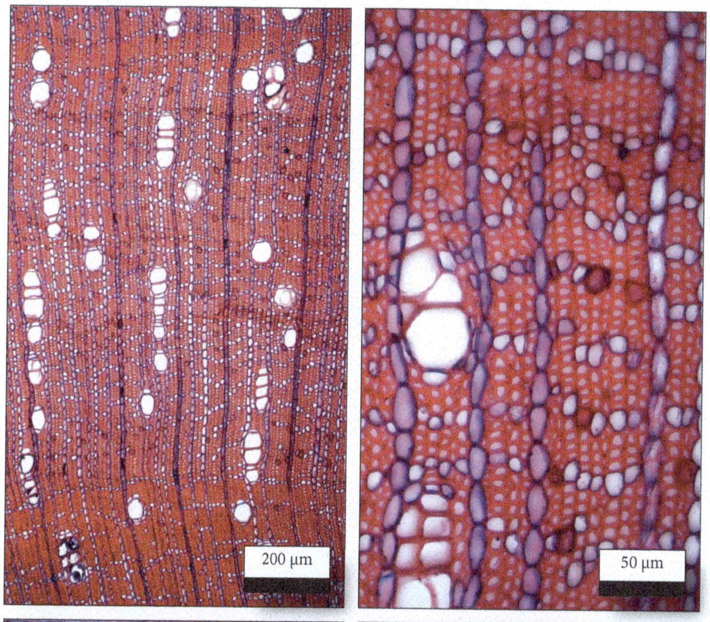

STEM XYLEM ANATOMY

Transverse section
Growth rings boundaries indistinct or absent.
Wood diffuse-porous.
Vessels solitary and in radial multiples of 2 to 6 elements.
Tangential diameter of vessels medium 60-74-90 µm, 30-36-43 vessels/mm^2.
Gums and/or other deposits in heartwood vessels.
Fibres thin- to thick-walled.
Apotracheal parenchyma diffuse-in-aggregates, paratracheal parenchyma scanty.
Rays 20-25-28/mm.

Radial section
Simple perforation plates.
Intervessel pits circular and alternate, 4-7 µm in diameter.
Vessel-ray pits with distinct borders; similar to intervessel pits in size and shape throughout the ray cell.
Vessel element length 234-280-362 µm.
Fibres with simple to minutely bordered pits (libriform fibers).
Heterogeneous rays, body ray cells procumbent with 1 to over 4 rows of upright and square marginal cells.
Prismatic crystal in chambered axial parenchyma cells, and in procumbent ray cells.

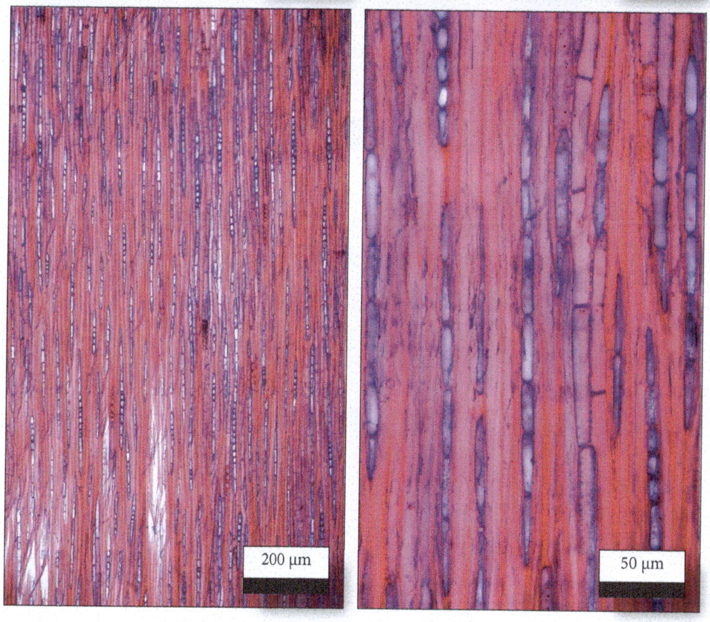

Tangential section
Rays exclusively uniseriate.
Ray height 260-505-920 µm.
Three to 8 cells per parenchyma strand.

Diospyros microrhombus Hiern

EN: Ebony FR: Bois d'ébène ML: Hazomainty

PLANT DESCRIPTION

Shrub to tree 4-5 m high.
Endemic to Madagascar. Non-exploited species.

DISTRIBUTION

Elevation: 0-500 m a.s.l.
Vegetation formation: Forest
Bioclimate: Humid
Protected areas: Andohahela and Manombo
New protected areas: Mahabo-Mananivo

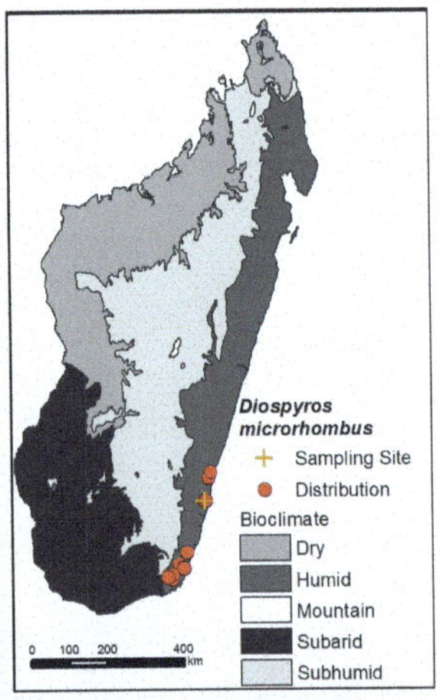

Slides made from material collected in Mahabo-Mananivo.

PITH AND PRIMARY XYLEM ANATOMY

Pith: Shape of pith round. Cells dimorphic. Pits in transverse cell walls. Pith not visible in polarized light.

Primary xylem: Vascular bundles separated. Vessels solitary and in short radial multiples.

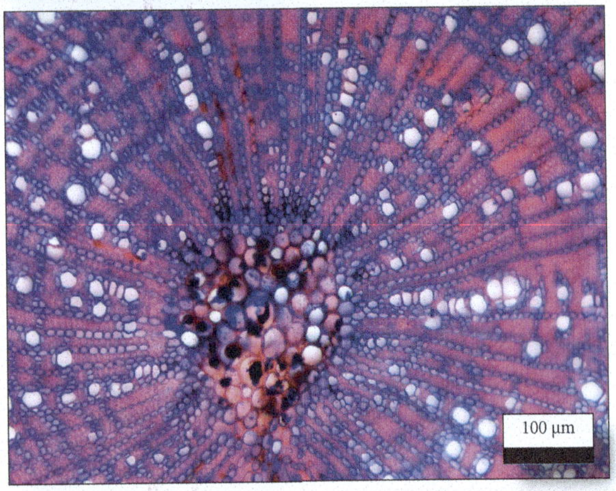

BRANCH XYLEM ANATOMY

Description: Growth rings boundaries indistinct or absent. Tangential diameter of vessels 50-63-80 µm, 22-34-46 vessels/mm^2. Vessel element length 200-269-350 µm. Intervessel pits small, 4-7 µm in diameter. Fibers very thick-walled. Axial parenchyma diffuse and diffuse-in-aggregates. Rays exclusively uniseriate, ray height 140-413-1010 µm, 16-18-21 rays/mm.

Anatomical differences to stem xylem: Mean tangential diameter of vessel lumina is similar in the branch and in the stem (50-63-80/40-64-80 µm), while vessel density is higher in the branch (22-34-46/12-20-26 vessels/mm^2). Ray height and number are lower in the branch (140-413-1010/160-510-980 µm; 16-18-21/30-35-38 rays/mm).

BARK ANATOMY

Branch bark: Collapsed and non-collapsed sieve tubes present in the phloem. The cortex is characterized by a band of sclereids and irregularly distributed parenchyma cells with dark staining substances in their lumina.

Stem bark: The phloem is characterized by collapsed and non-collapsed sieve tubes and tangentially distributed groups of sclereids. Rays and Axial parenchyma dilatation present. Prismatic crystals present in the sclereids. The cortex is characterized by groups of sclereids irregularly distributed. Layered phellem. The phellem is homogeneous with dark staining substances and it consists of regularly arranged rectangular cells, indistinct in polarized light.

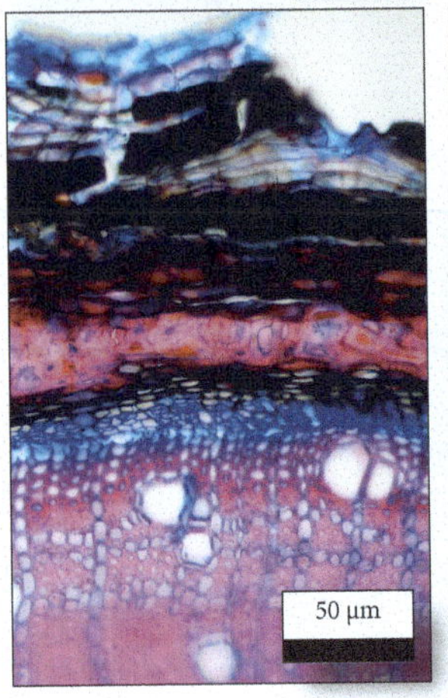

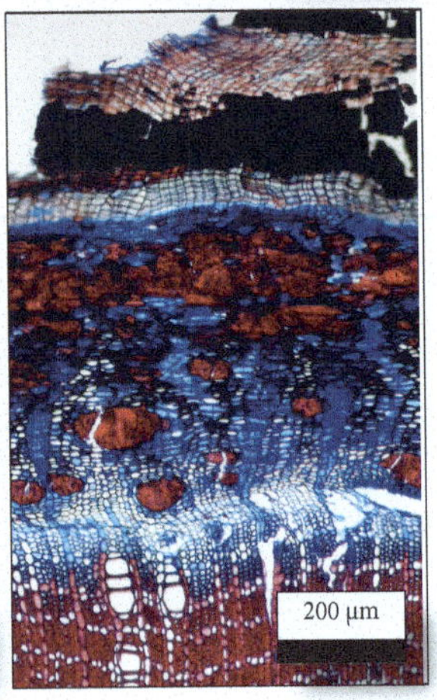

IAWA feature numbers

2 5 9 9.1 13 22 25 30 41 47 53 58 61 66 70 77 79 86 92 96.1 102.4 107 108 116 136 138 142 180 189 190 192 196

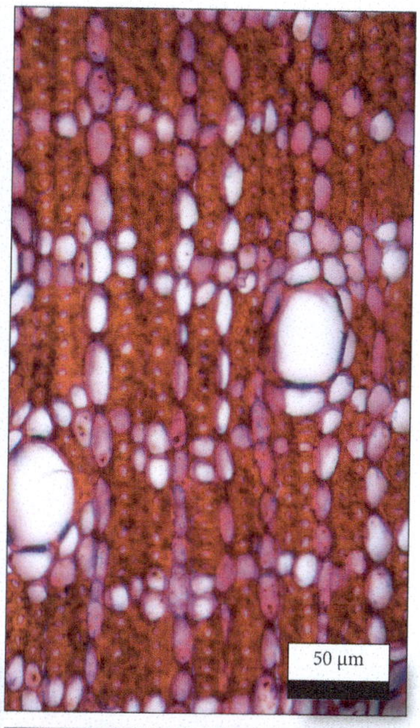

STEM XYLEM ANATOMY

Transverse section

Growth ring boundaries absent.
Wood diffuse-porous.
Vessels solitary and in short radial multiples. Tangential diameter of vessels medium 40-64-80 µm, 16-20-26 vessels/mm^2.
Gums and/or other deposits in heartwood vessels.
Fibers very thick-walled.
Apotracheal parenchyma diffuse-in-aggregates and in narrow lines up to 3 cells. Paratracheal parenchyma vasicentric.
Rays 30-35-38/mm.

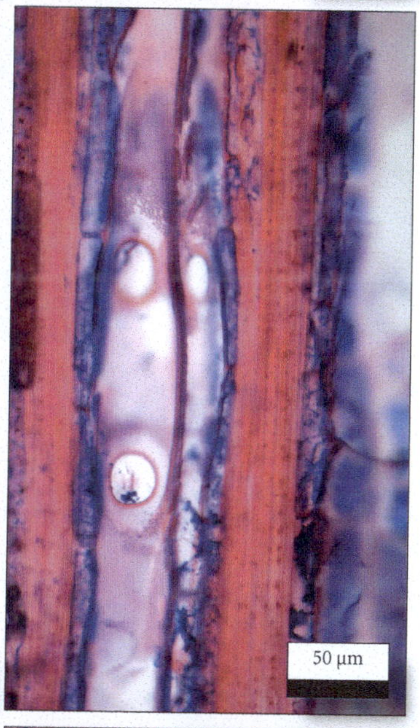

Radial section

Simple perforation plates.
Intervessel pits circular and alternate, 4-7 µm in diameter.
Vessel-ray pits with distinct borders; similar to intervessel pits in size and shape throughout the ray cell.
Vessel element length 210-385-520 µm.
Fibres with simple to minutely bordered pits (libriform fibers).
Heterogeneous rays, body ray cells procumbent with 2 to over 4 rows of upright and square marginal cells.
Prismatic crystals in procumbent ray cells and in chambered axial parenchyma cells.

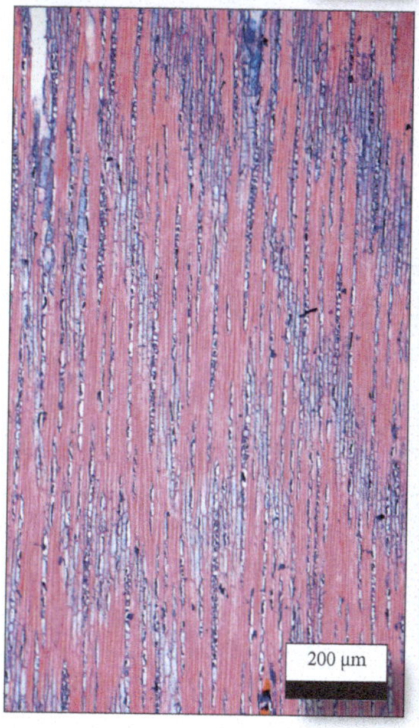

Tangential section

Rays uni- and biseriate, predominantly uniseriate.
Ray height 160-510-980 µm.
Two to 4 cells per parenchyma strand.

Diospyros occlusa H. Perrier

Synonymy: *Maba magnifolia* H. Perrier

EN: Ebony FR: Bois d'ébène ML: Hazomainty

PLANT DESCRIPTION

Shrub to tree | 7-25 m high | 12-30 cm DBH

Endemic to Madagascar. Exploited species.

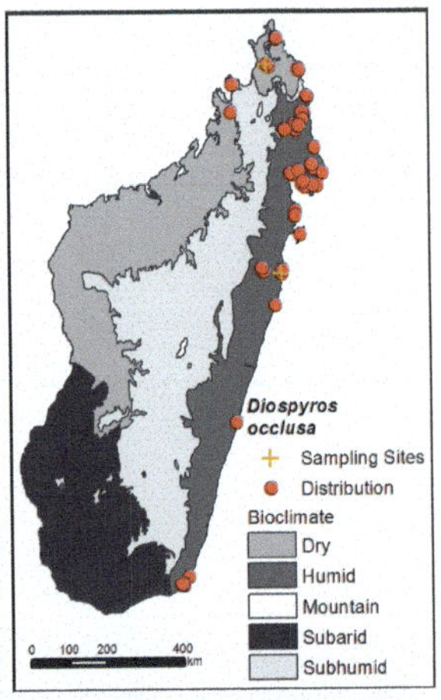

DISTRIBUTION

Elevation: 0-500 m a.s.l.

Vegetation formation: Forest

Bioclimate: Humid and sub-humid

Protected areas: Ankarana, Lokobe, Mananara-Nord, Marojejy, Masoala, Nosy Mangabe and Zahamena

New protected areas: Ile Ste Marie and Montagne des Français

Slides made from material collected in Ankarana.

PITH AND PRIMARY XYLEM ANATOMY

Pith: Shape of pith oval. Thick-walled parenchyma cells present. Cells dimorphic. Prismatic crystals present. Pits in transverse cell walls. Pith not visible in polarized light.

Primary xylem: Vascular bundles separated. Vessels solitary and in long radial multiples.

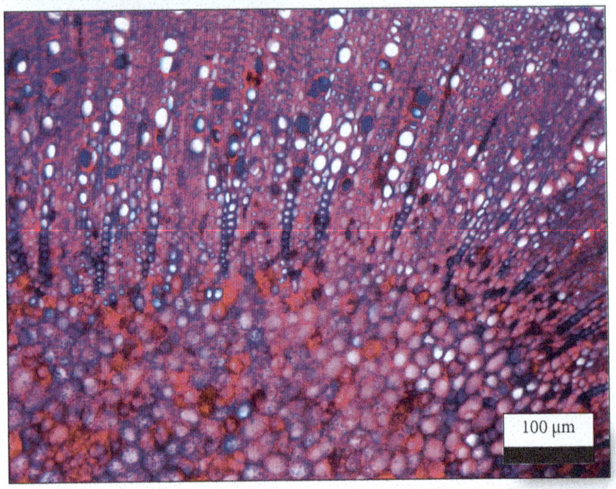

BRANCH XYLEM ANATOMY

Description: Growth rings boundaries indistinct or absent. Tangential diameter of vessels 10-17-25 µm, 127-152-177 vessels/mm². Vessel element length 160-318-460 µm. Intervessel pits small, 4-7 µm in diameter. Fibres very thick-walled. Axial parenchyma diffuse and diffuse-in-aggregates. Rays exclusively uniseriate, ray height 150-295-450 µm, 19-21-24 rays/mm.

Anatomical differences to stem xylem: Mean tangential diameter of vessel lumina is smaller in the branch (10-17-25 µm/17-25-38 µm). Vessel density is slightly higher in the branch than in the stem (127-152-177/120-143-160 vessels/mm²). Rays are higher in the branch than in the stem (150-295-450 µm/50-230-400 µm), while ray number is similar (19-21-24/18-20-21 rays/mm).

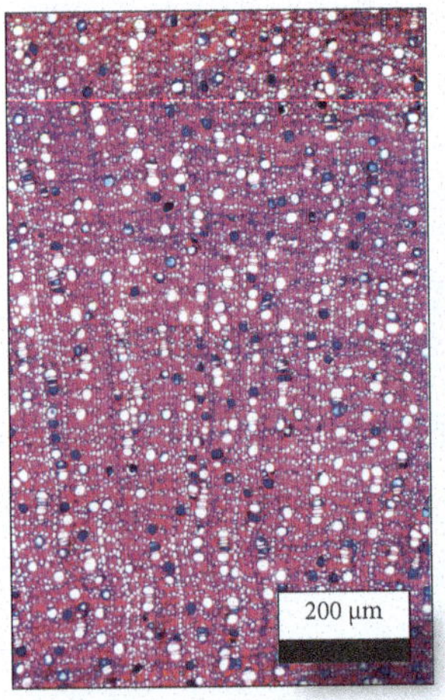

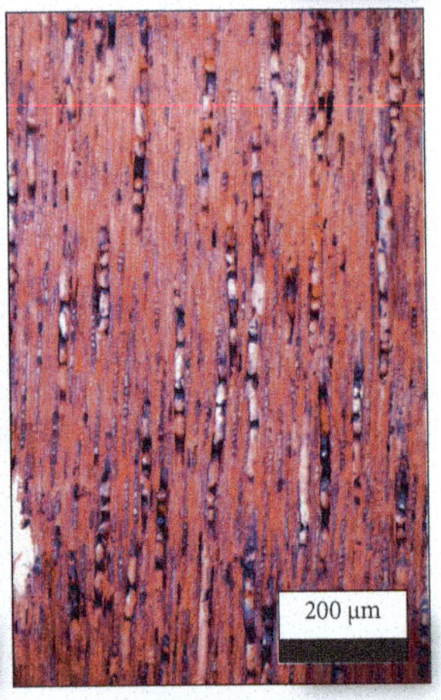

BARK ANATOMY

Branch bark: Collapsed and non-collapsed sieve tubes present in the phloem. Some rays become dilated. The cortex is characterized by an initial band of fibers and sclereids, and by single sclereids irregularly distributed in blue-stained parenchyma cells.

Stem bark: The phloem consists of alternating tangentially distributed groups of sclereids, and collapsed and non-collapsed sieve tubes. Rays and parenchyma cells dilated. The phellem is homogeneous and it consists of regularly arranged rectangular cells, indistinct in polarized light.

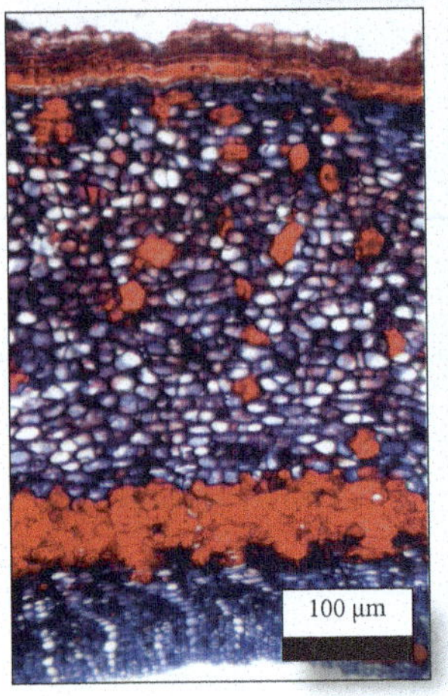

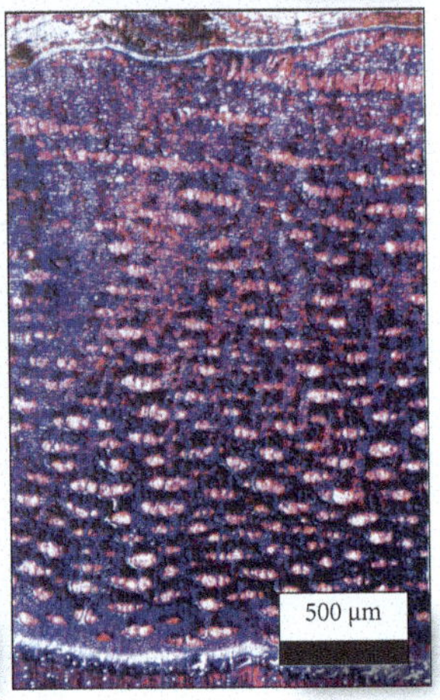

IAWA feature numbers
2 5 9.1 10 13 22 25 30 40 50 58 61 66 70 76 77 78 92
96.1 102.3 107 116 136 142 180 189 190 192 196

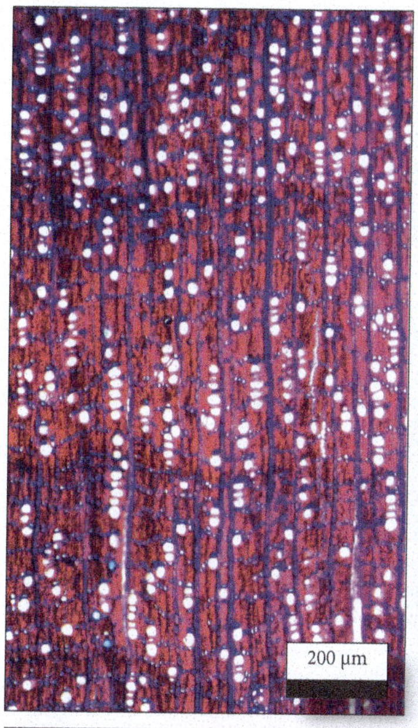

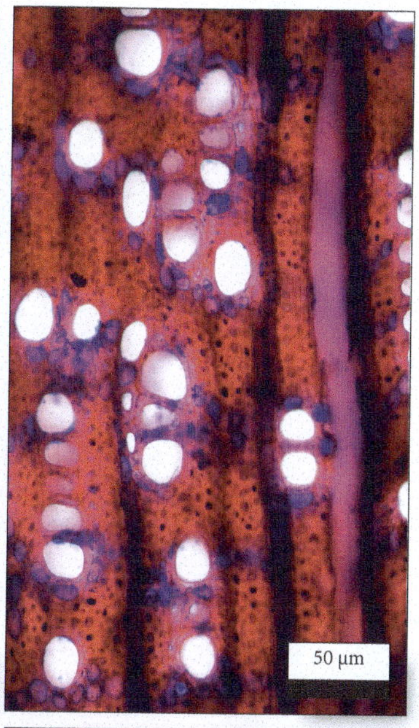

STEM XYLEM ANATOMY

Transverse section

Growth rings boundaries indistinct or absent.
Wood diffuse-porous.
Vessels solitary and in radial multiples 2 to 5 elements.
Tangential diameter of vessels small 17-25-38 µm, 120-143-160 vessels/mm^2.
Gums and/or other deposits in heartwood vessels.
Fibers very thick-walled.
Apotracheal parenchyma diffuse, and diffuse-in-aggregates. Paratracheal parenchyma scanty.
Rays 18-20-21/mm.

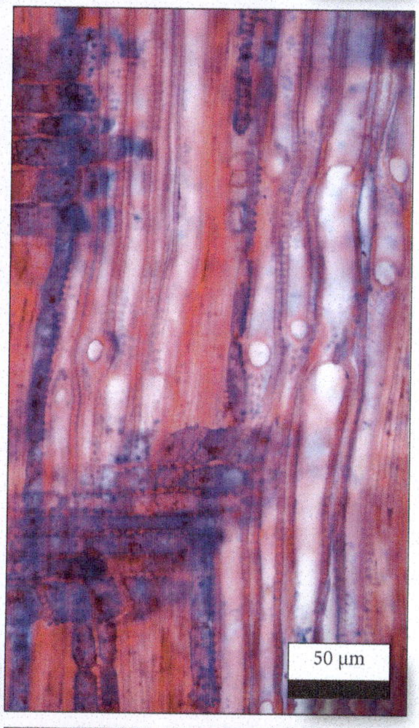

Radial section

Simple perforation plates.
Intervessel pits circular and alternate, 4-7 µm in diameter.
Vessel-ray pits with distinct borders; similar to intervessel pits in size and shape throughout the ray cell.
Vessel element length 200-294-440 µm.
Fibres with simple to minutely bordered pits (libriform fibers). Non-septate fibres present.
Heterogeneous rays, body ray cells procumbent with 2 to 4 rows of upright marginal cells.
Prismatic crystal in chambered axial parenchyma cells.

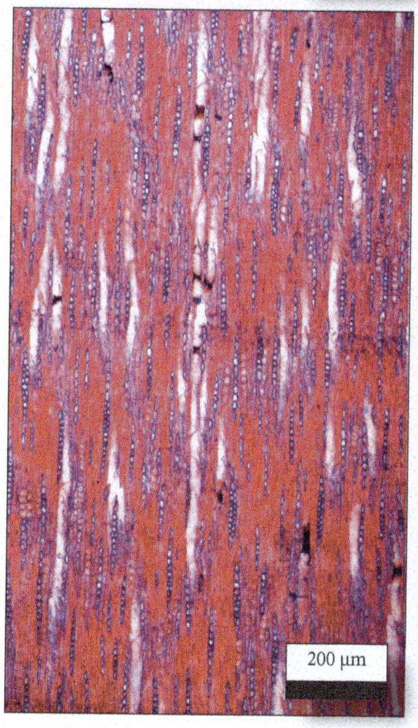

Tangential section

Rays exclusively uniseriate.
Ray height 50-230-400 µm.
Two to 4 cells per parenchyma strand.

Diospyros olacinoides (H. Perrier) G.E. Schatz and Lowry

Synonymy: *Maba gneissicola* H. Perrier; *Maba olacinoides* H. Perrier.

EN: Ebony FR: Bois d'ébène ML: Hazomainty

PLANT DESCRIPTION

Shrub to tree | 6-10 m | 13 cm DBH
Endemic to Madagascar. Non-exploited species.

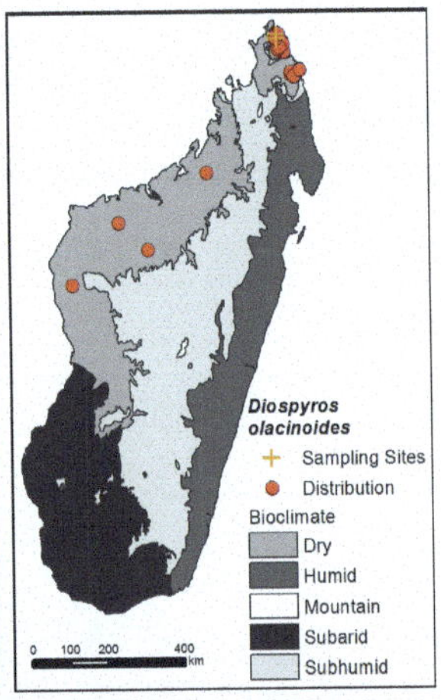

DISTRIBUTION

Elevation: 0-500 m a.s.l.
Vegetation formation: Forest
Bioclimate: Humid
Protected areas: Analamerana and Namoroka
New protected areas: Daraina complex, Montagne des Français, Sahafary and Orangea/Rigny

Slides made from material collected in Orangea/Rigny.

PITH AND PRIMARY XYLEM ANATOMY

Pith: Shape of pith oval. Thick-walled parenchyma cells present. Cells dimorphic. Prismatic crystals present. Pits in transverse cell walls. Pith not visible in polarized light.

Primary xylem: Vascular bundles separated. Vessels solitary and in long radial multiples.

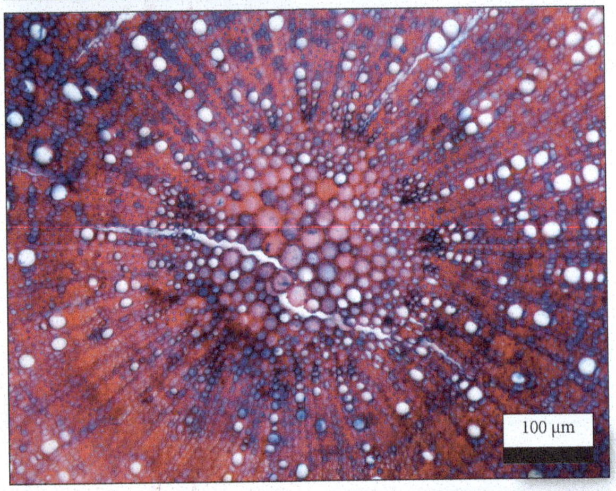

BRANCH XYLEM ANATOMY

Description: Growth rings boundaries indistinct or absent. Tangential diameter of vessel 42-48-55 µm, 212-229-250 vessels/mm^2. Vessel element length 210-326-390 µm. Intervessel pits small, 4-7 µm in diameter. Fibres very thick-walled. Axial parenchyma diffuse, and diffuse-in-aggregates. Rays exclusively uniseriate, ray high 120-369-870 µm, 22-26-30 rays/mm.

Anatomical differences to stem xylem: Mean tangential diameter of vessel lumina is slightly smaller in the branch (42-48-55/40-53-78 µm). Vessel density is higher in the branch than in the stem (212-229-250/111-131-150 vessels/mm^2). Rays are slightly higher in the branch than in the stem (120-369-870/190-349-780 µm), ray number is higher in the branch than in the stem (22-26-30/18-22-25 rays/mm).

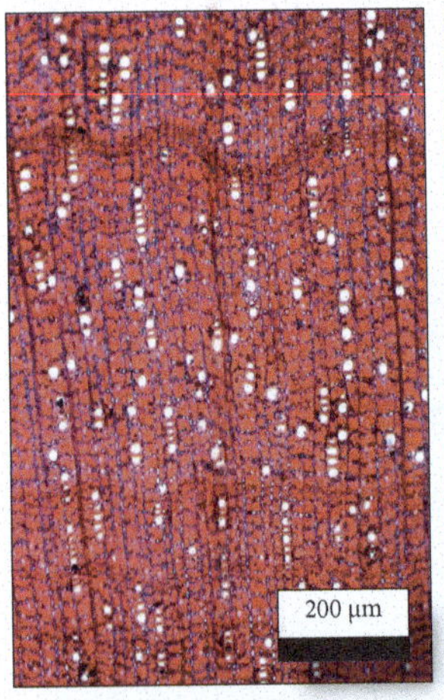

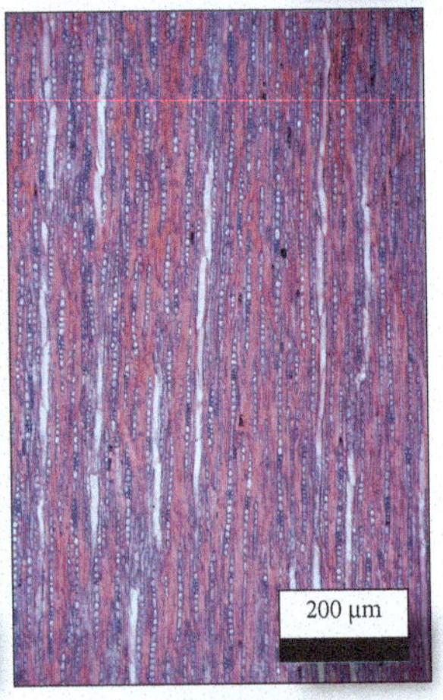

BARK ANATOMY

Branch bark: Collapsed and non-collapsed sieve tubes with groups of sclereids present in the phloem. The cortex consists of a band of sclereids and irregularly shaped blue-stained parenchyma cells.

Stem bark: The phloem consists of alternating collapsed and non-collapsed sieve tubes, and irregularly distributed groups of sclereids. Prismatic crystals present in the sclereids. The phellem is homogeneous and it consists on regularly arranged of rectangular cells, indistinct in polarized light.

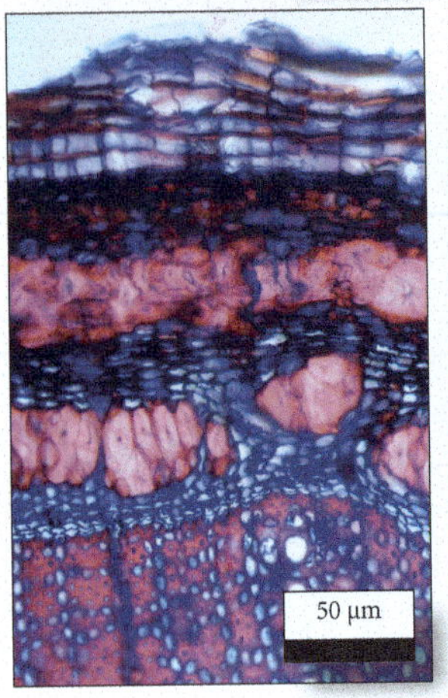

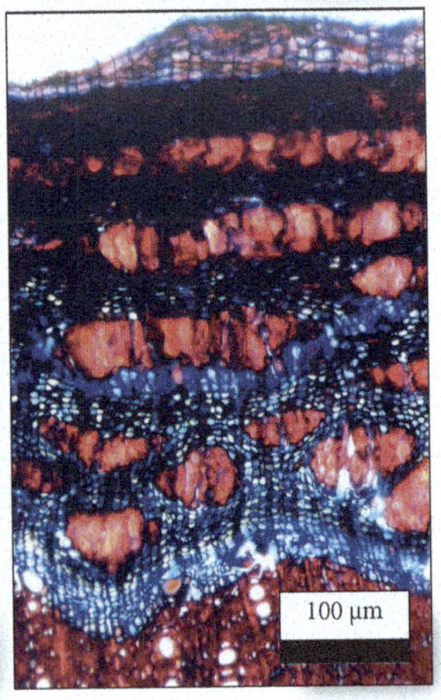

IAWA feature numbers
2 5 9.1 10 13 22 25 30 40 41 50 52 58 61 66 70 78 79 92
96.1 100 102.3 107 116 136 142 180 189 192 196

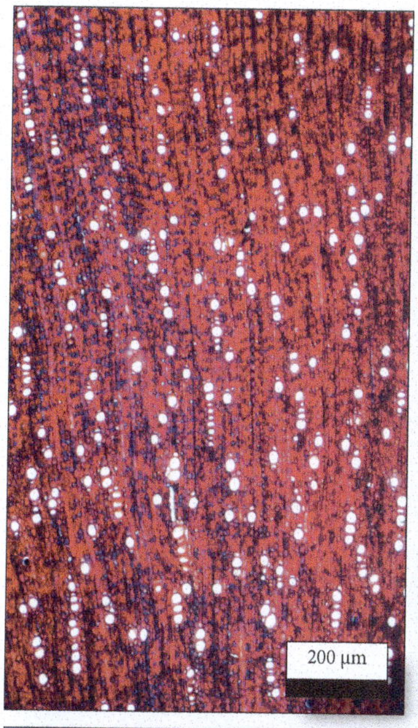

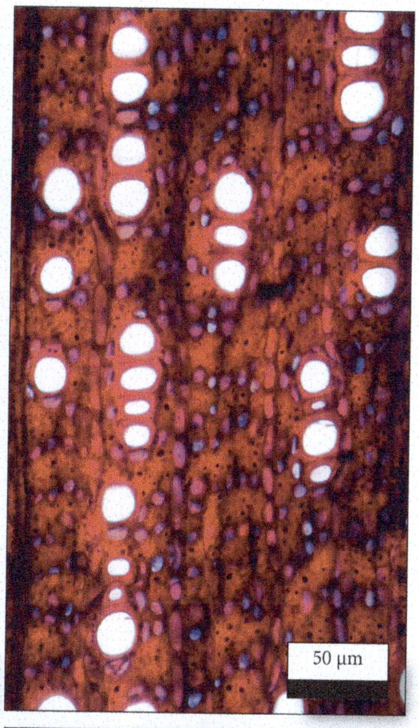

STEM XYLEM ANATOMY

Transverse section

Growth rings boundaries indistinct or absent.
Wood diffuse-porous.
Vessels solitary and in radial multiples of 2 to 6 elements.
Tangential diameter of vessels small 40-53-78 µm, 111-131-150 vessels/mm^2.
Fibers very thick-walled.
Dark staining substances present in heartwood vessel lumina.
Apotracheal parenchyma diffuse-in-aggregate.
Paratracheal parenchyma scanty to vasicentric.
Rays 18-22-25/mm.

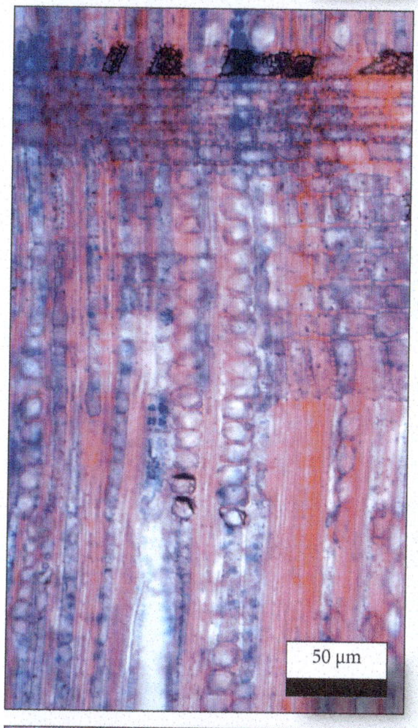

Radial section

Simple perforation plates.
Intervessel pits circular and alternate, 4-7 µm in diameter.
Vessel-ray pits with distinct borders; similar to intervessel pits in size and shape throughout the ray cell.
Vessel element length 190-292-400 µm.
Fibres with simple to minutely bordered pits (libriform fibers).
Heterogeneous rays, body ray cells procumbent with 2 to 4 rows of upright marginal cells.
Prismatic crystal in chambered axial parenchyma cells.

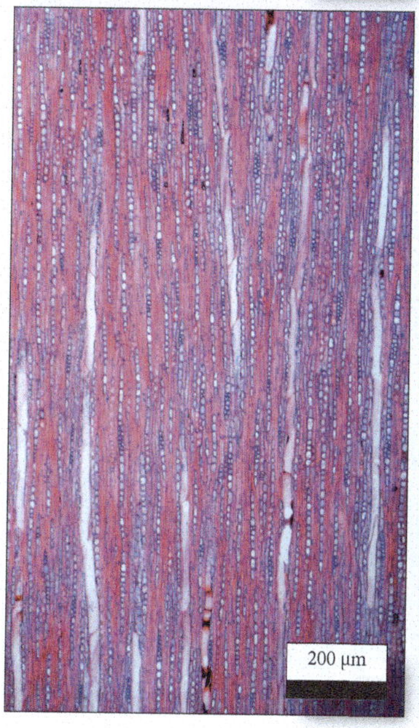

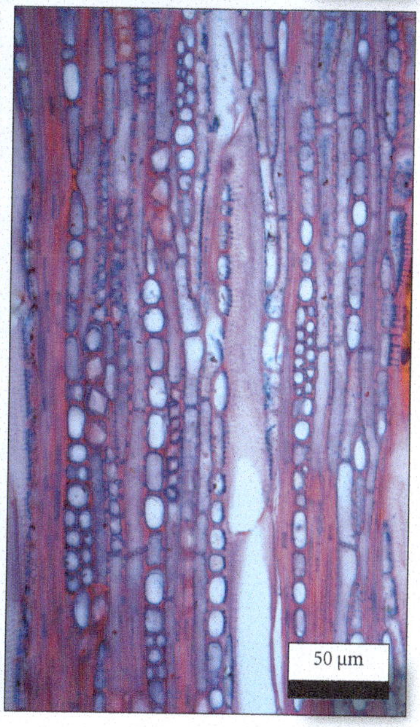

Tangential section

Rays uni- and biseriate, predominantly biseriate.
Rays with multiseriate portion as wide as uniseriate portions.
Ray height 190-349-780 µm.
Two to 4 cells per parenchyma strand.

Diospyros parifolia (Schltr.) Bakh.

EN: Ebony FR: Bois d'ébène ML: Hazomainty

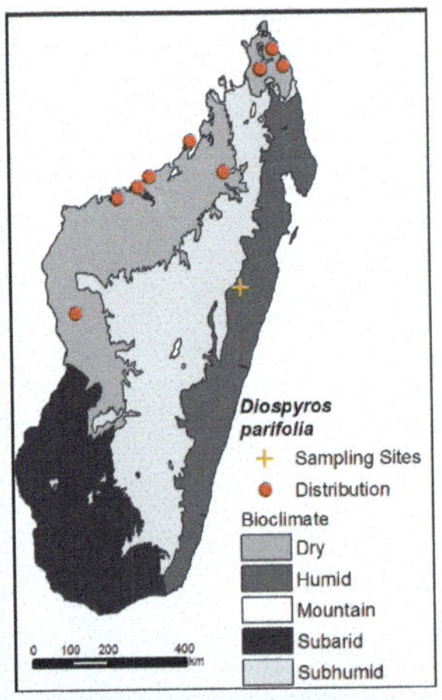

PLANT DESCRIPTION

Shrub to tree | 3-12 m high | 9-15 cm DBH
Endemic to Madagascar. Non-exploited species.

DISTRIBUTION

Elevation: 0-500 m a.s.l.
Vegetation formation: Forest
Bioclimate: Dry
Protected areas: -
New protected areas: Daraina complex, Orangea/Rigny and Didy

Slides made from material collected in Didy.

PITH AND PRIMARY XYLEM ANATOMY

Pith: Shape of pith round. Thick-walled parenchyma cells present. Cells dimorphic. Pits in transverse cell walls. Pith visible in polarized light.

Primary xylem: Vascular bundles separated. Vessels solitary and in long radial multiples.

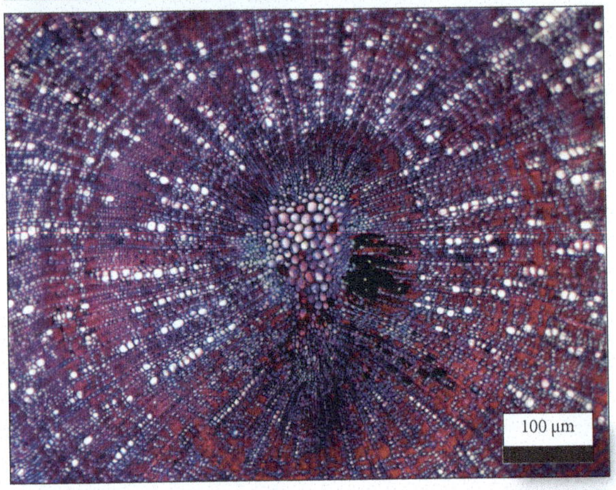

BRANCH XYLEM ANATOMY

Description: Growth rings boundaries indistinct or absent. Tangential diameter of vessels 60-75-100 µm, 35-46-50 vessels/mm^2. Vessel element length 200-394-600 µm. Intervessel pits small, 4-7 µm in diameter. Fibres thick- to very thick-walled. Axial parenchyma diffuse-in-aggregates, and in narrow lines up to 3 cells wide. Rays exclusively uniseriate, 210-403-790 µm high, 26-35-41 rays/mm.

Anatomical differences to stem xylem: Mean tangential diameter of vessel lumina is slightly smaller in the branch than in the stem (60-75-100 µm/70-86-110 µm), while vessel density is higher in the branch than in the stem (35-46-50/13-22-29 vessels/mm^2). Rays are higher in the branch (210-403-790/150-374-710 µm), while ray number is slightly higher in the stem (26-35-41/18-25-31 rays/mm).

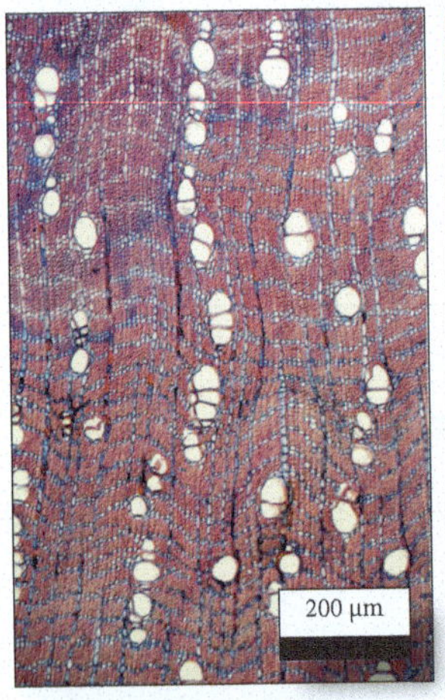

BARK ANATOMY

Branch bark: Collapsed and non-collapsed sieve tubes present in the phloem. Some rays become dilated. The cortex is characterized by a band of sclereids and irregular blue-stained parenchyma cells.

Stem bark: The phloem consists of alternating groups of tangentially distributed sclereids and collapsed and non-collapsed sieve tubes. Prismatic crystals present in the sclereids. The phellem cells are homogeneous with dark staining substances and it consists of arranged rectangular cells, indistinct in polarized light.

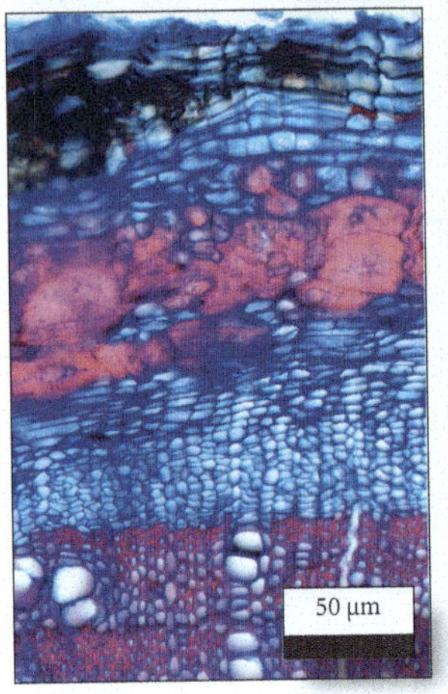

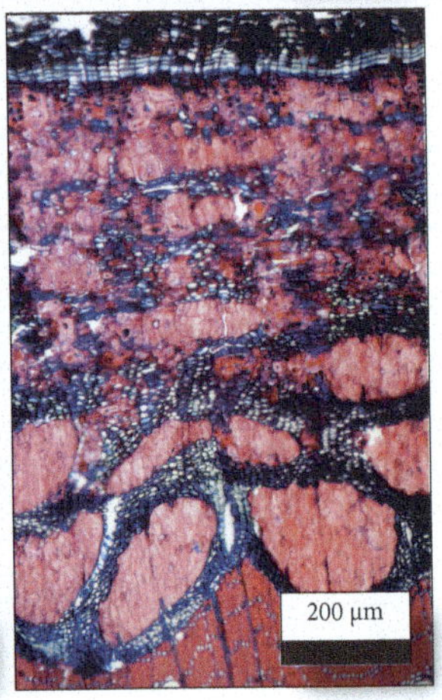

IAWA feature numbers
2 5 9.1 10 13 22 25 30 41 48 52 58 61 66 70 77 78 86 92 96.1 102.3 107 108 116 136 138 142 180 189 190 192 196

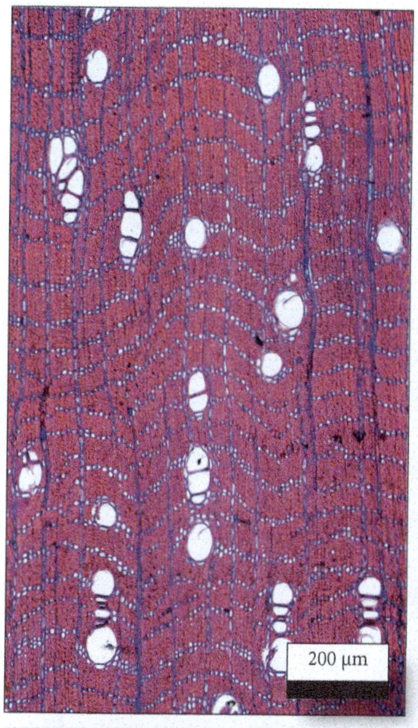

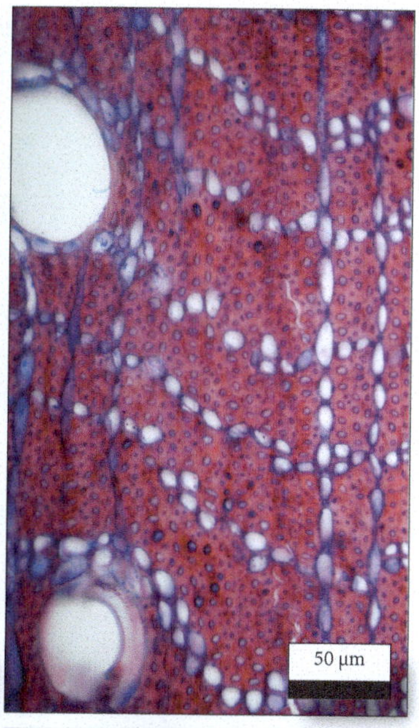

STEM XYLEM ANATOMY

Transverse section
Growth ring boundaries absent.
Wood diffuse-porous.
Vessels solitary and in radial multiples of 2 to 4 elements.
Tangential diameter of vessels 70-86-110 µm, 13-22-29 vessels/mm^2.
Dark staining substances present in heartwood vessel lumina.
Fibers very thick-walled.
Apotracheal parenchyma diffuse-in-aggregates, and in narrow bands up to three cells wide.
Rays 18-25-31/mm.

Radial section
Simple perforation plates.
Intervessel pits circular and alternate, 4-7 µm in diameter.
Vessel-ray pits with distinct borders; similar to intervessel pits in size and shape throughout the ray cell.
Vessel element length 260-325-400 µm.
Fibres with simple to minutely bordered pits (libriform fibers).
Heterogeneous rays, body ray cells procumbent with 2 to over 4 rows of upright marginal cells.
Prismatic crystals in procumbent ray cells and in chambered axial parenchyma cells.

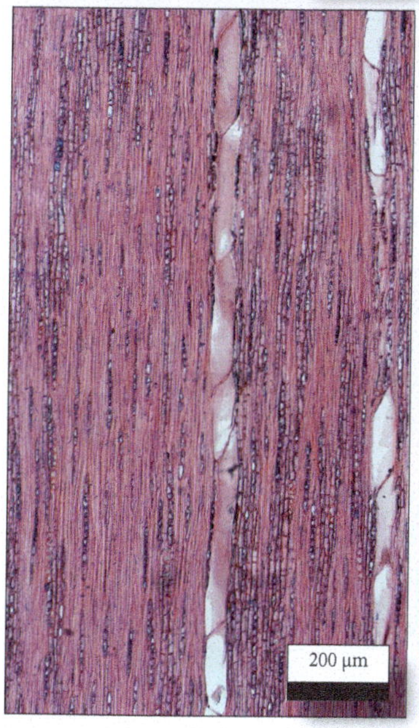

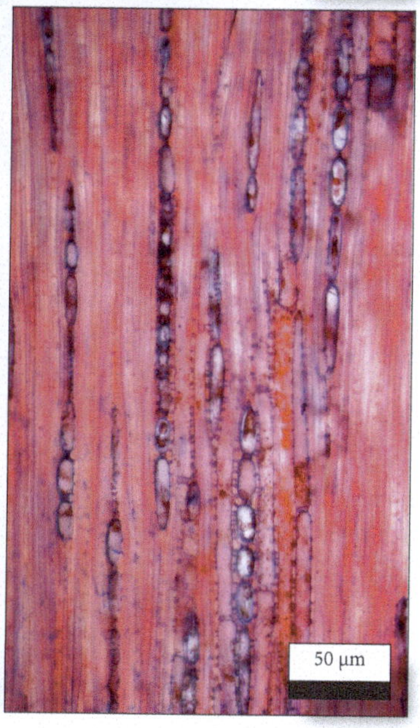

Tangential section
Rays uni- and biseriate, predominantly uniseriate.
Ray height 150-374-710 µm.
Three to 4 cells per parenchyma strand.

Diospyros perrieri Jum.

EN: Ebony FR: Bois d'ébène ML: Hazomainty

PLANT DESCRIPTION

Tree | 5-25 m high
Endemic to Madagascar. Exploited species.

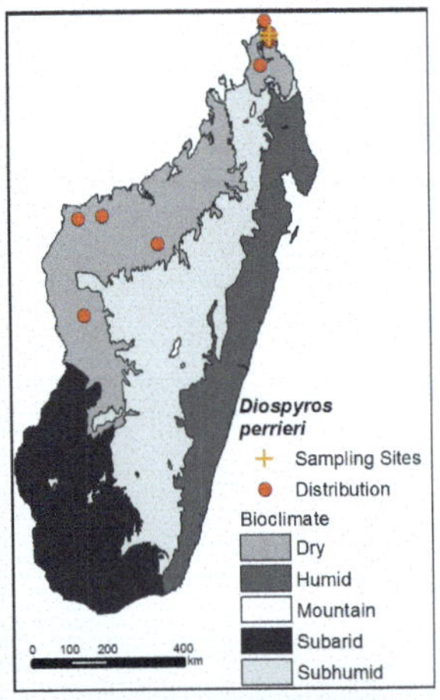

DISTRIBUTION

Elevation: 0-500 m a.s.l.
Vegetation formation: Forest and woodland
Bioclimate: Dry
Protected areas: Ankarana, Bemaraha and Namoroka
New protected areas: Orangea/Rigny and Montagne des Français

Slides made from material collected in Montagne des Français.

PITH AND PRIMARY XYLEM ANATOMY

Pith: Shape of pith round. Thick-walled parenchyma cells present. Cells dimorphic. Prismatic crystals present. Pits in transverse cell walls. Pith visible in polarized light.

Primary xylem: Vascular bundles not distinct. Vessels solitary and in short radial multiples.

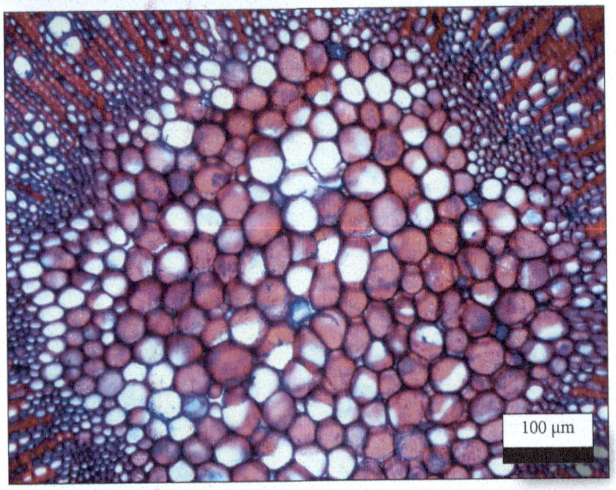

BRANCH XYLEM ANATOMY

Description: Growth rings boundaries indistinct or absent. Tangential diameter of vessels 42-56-70 µm, 45-60-70 vessels/mm^2. Vessel element length 200-294-400 µm. Intervessel pits small, 4-7 µm in diameter. Fibres thick- to very thick-walled. Axial parenchyma diffuse-in-aggregates, and in narrow bands up to three cells wide. Rays exclusively uniseriate, ray height 140-404-800 µm, 18-24-32 rays/mm.

Anatomical differences to stem xylem: Mean tangential diameter of vessel lumina is slightly smaller in the branch than in the stem (42-56-70/40-64-80 µm), while vessel density is higher in the branch (45-60-70/15-26-30 vessels/mm^2). Ray height and ray number are almost similar in branch and stem (140-404-800/120-399-940 µm; 18-24-32/18-22-26 rays/mm).

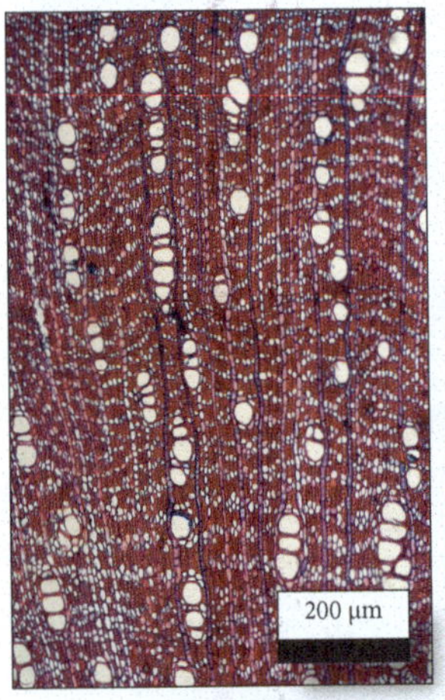

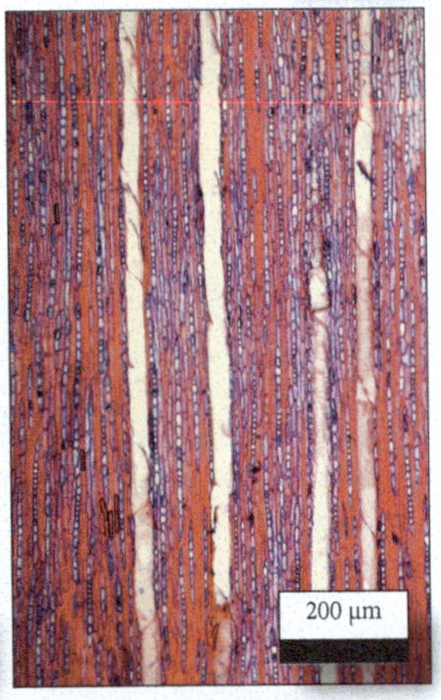

BARK ANATOMY

Branch bark: Collapsed and non-collapsed sieve tubes present in the phloem. Tangentially distributed groups of sclereids in the phloem. The cortex is characterized by a band of sclereids and irregularly shaped blue-stained parenchyma cells.

Stem bark: The phloem consists of alternating tangentially distributed groups of sclereids, and collapsed and non-collapsed sieve tubes. Prismatic crystals in the sclereids. The phellem cells are homogeneous with dark staining substances. Phellem cells rectangular, indistinct in polarized light.

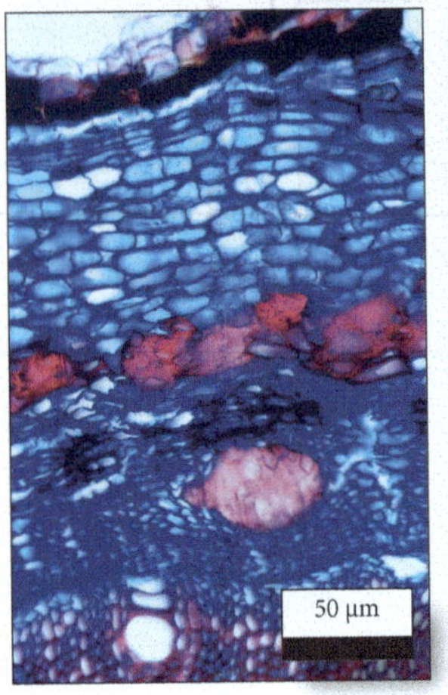

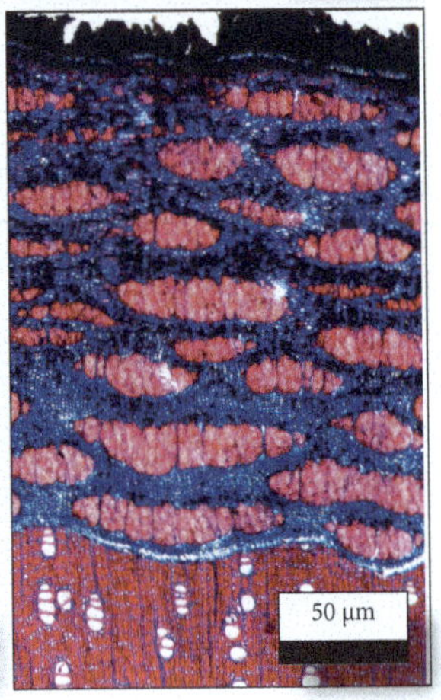

IAWA feature numbers
2 5 10 13 22 25 30 41 48 52 58 61 66 70 77 78 86 92
93 96.1 102.3 107 108 116 136 138 142 180 189 192 196

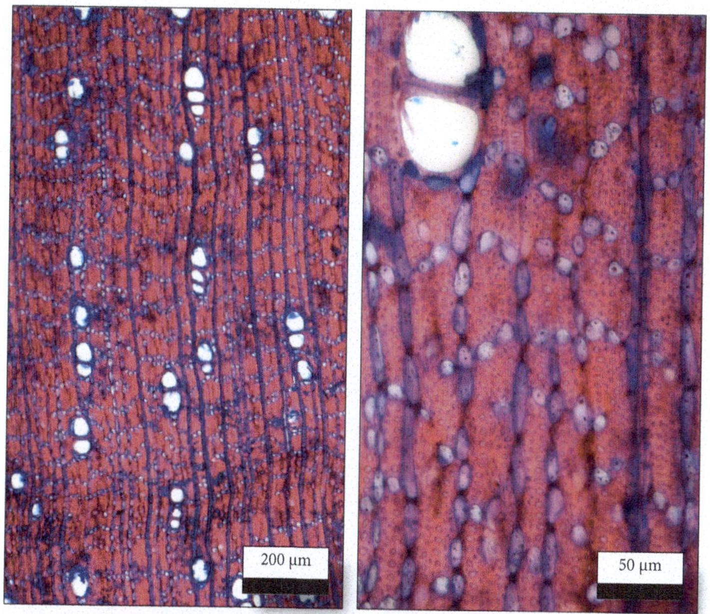

STEM XYLEM ANATOMY

Transverse section

Growth rings boundaries indistinct or absent.
Wood diffuse-porous.
Vessels solitary and in radial multiples of 2 to 6 elements.
Tangential diameter of vessels medium 40-64-80 μm, 15-26-30 vessels/mm^2.
Dark staining substances present in heartwood vessel lumina.
Fibers very thick-walled.
Apotracheal parenchyma diffuse-in-aggregates, and in narrow bands up to three cells wide.
Paratracheal parenchyma scanty.
Rays 18-22-26/mm.

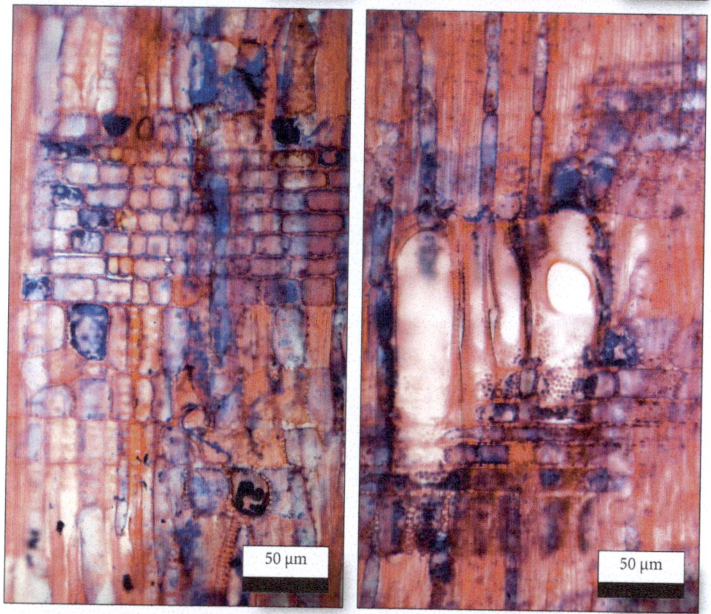

Radial section

Simple perforation plates.
Intervessel pits circular and alternate, 4-7 μm in diameter.
Vessel-ray pits with distinct borders; similar to intervessel pits in size and shape throughout the ray cell.
Vessel element length 80-313-450 μm.
Fibres with simple to minutely bordered pits (libriform fibers).
Heterogeneous rays, body ray cells procumbent with 2 to over 4 rows of upright marginal cells.
Prismatic crystals in procumbent ray cells.

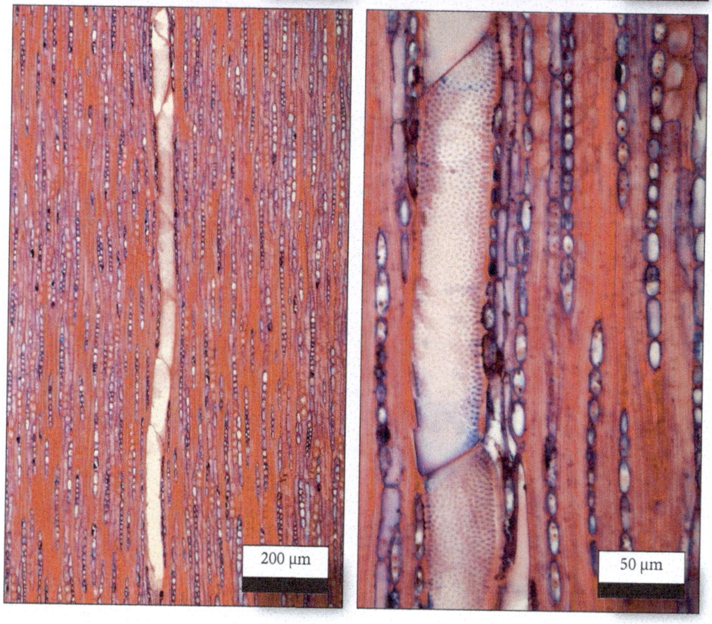

Tangential section

Rays uni- and biseriate, predominantly uniseriate.
Ray height 120-399-940 μm.
Two to 8 cells per parenchyma strand.
Prismatic crystal in chambered axial parenchyma cells.

Diospyros pervilleana (Baill.) G.E. Schatz & Lowry

Synonymy: *Maba diffusa* Hiern; *Maba pervilleana* (Baill.) H. Perrier; *Olax pervilleana* Baill.

EN: Ebony FR: Bois d'ébène ML: Hazomainty

PLANT DESCRIPTION

Tree | 4-9 m high | 4-10 cm DBH
Endemic to Madagascar. Non-exploited species.

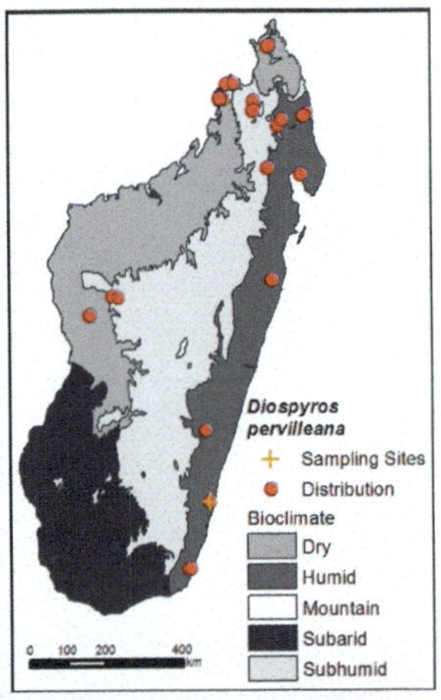

DISTRIBUTION

Elevation: 0-2000 m a.s.l.
Vegetation formation: Forest
Bioclimate: Dry, humid and sub-humid
Protected areas: Ambohijanahary, Bemaraha, Lokobe, Montagne d'Ambre and Tsaratanana.
New protected areas: Mahabo-Mananivo.

Slides made from material collected in Mahabo-Mananivo.

PITH AND PRIMARY XYLEM ANATOMY

Pith: Shape of pith round. Almost all parenchyma cells thick-walled. Cells dimorphic. Pits in transverse cell walls. Pith visible in polarized light.

Primary xylem: Vascular bundles separated. Vessels solitary and in long radial multiples.

BRANCH XYLEM ANATOMY

Description: Growth rings boundaries indistinct or absent. Tangential diameter of vessels 10-14-20 µm, 26-33-41 vessels/mm^2. Vessel elements length 160-287-390 µm. Intervessel pits small, 4-7 µm in diameter. Fibres very thick-walled. Axial parenchyma diffuse-in-aggregates and scanty paratracheal. Rays exclusively uniseriate, ray height 150-265-420 µm, 30-35-42 rays/mm.

Anatomical differences to stem xylem: Mean tangential diameter of vessel lumina and vessel density are lower in the branch than in the stem (10-14-20/20-36-50 µm; 26-33-41/20-36-50 vessels/mm^2). Rays are lower in the branch (150-265-420/130-397-850 µm), while ray number is slightly higher in the branch than in the stem (30-35-42/22-26-35 rays/mm).

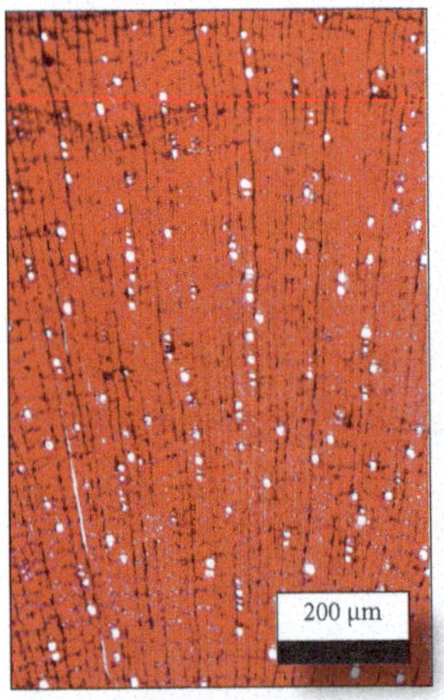

BARK ANATOMY

Branch bark: Collapsed and non-collapsed sieve tubes present in the phloem. The cortex is characterized by a band of sclereids and fibers, and irregularly shaped blue-stained parenchyma cells.

Stem bark: Collapsed and non-collapsed sieve tubes present. Single sclereids and tangentially arranged sclereid groups in the phloem. The phellem is homogeneous, phellem cells lumina filled with dark stained substances.

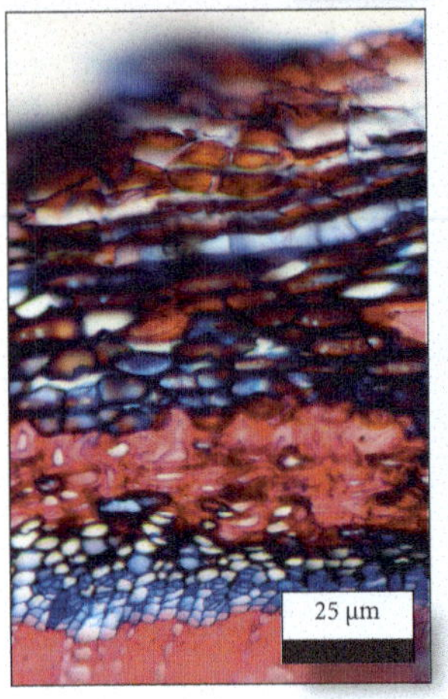

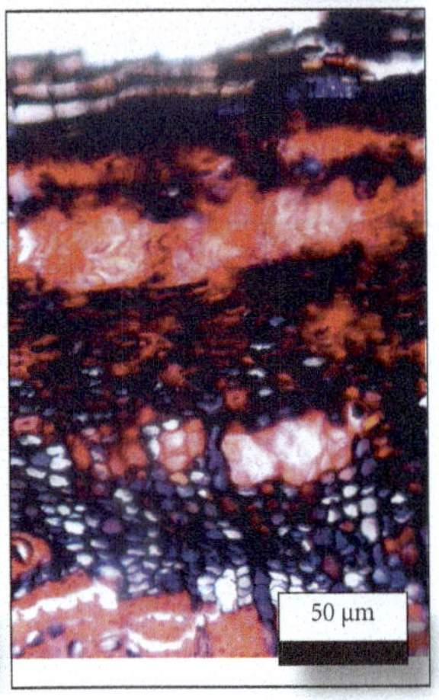

IAWA feature numbers
2 5 9.1 10 13 22 25 30 40 48 52 58 61 66 70 77 78 91 92 96 102.3 107 116 136 142 180 189 192 196

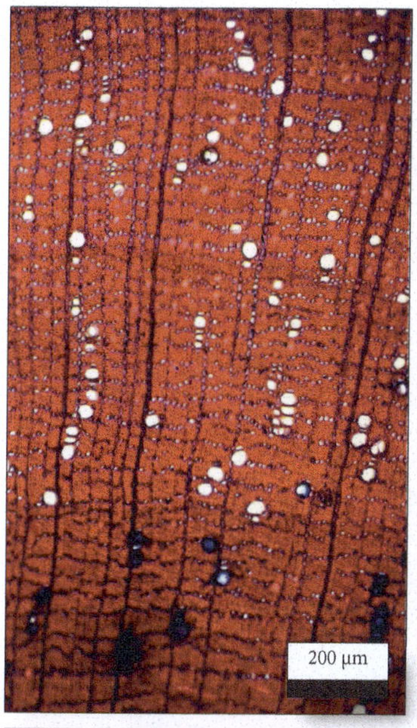

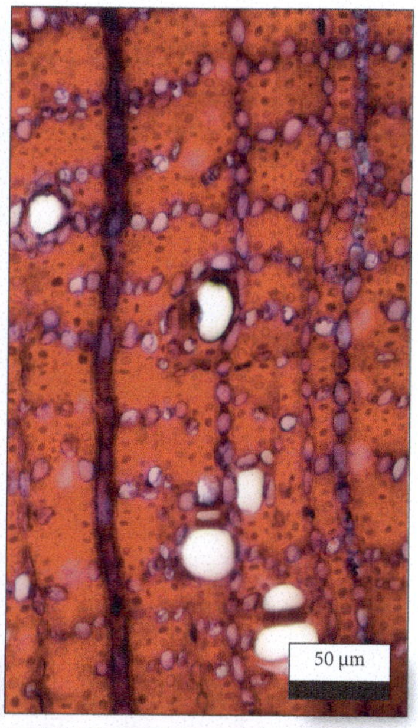

STEM XYLEM ANATOMY

Transverse section

Growth rings boundaries indistinct or absent.
Wood diffuse-porous.
Vessels solitary and in radial multiples of 2 to 6 elements. Tangential diameter of vessels 20-36-50 µm, 18-36-44 vessels/mm^2.
Dark staining substances present in heartwood vessel lumina.
Fibers very thick-walled.
Apotracheal parenchyma diffuse-in-aggregates, paratracheal parenchyma scanty.
Rays 22-26-35/mm.

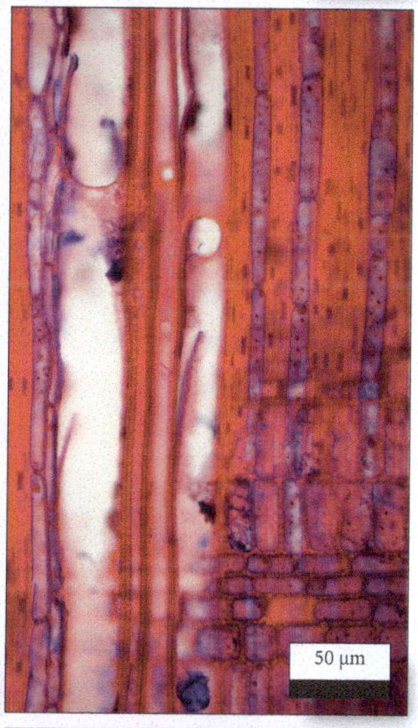

Radial section

Simple perforation plates.
Intervessel pits circular and alternate, 4-7 µm in diameter.
Vessel-ray pits with distinct borders; similar to intervessel pits in size and shape throughout the ray cell.
Vessel elements length 230-332-410 µm.
Fibres with simple to minutely bordered pits (libriform fibers).
Heterogeneous rays, body ray cells procumbent with 2 to 4 rows of upright marginal cells.
Prismatic crystal in chambered axial parenchyma cells.

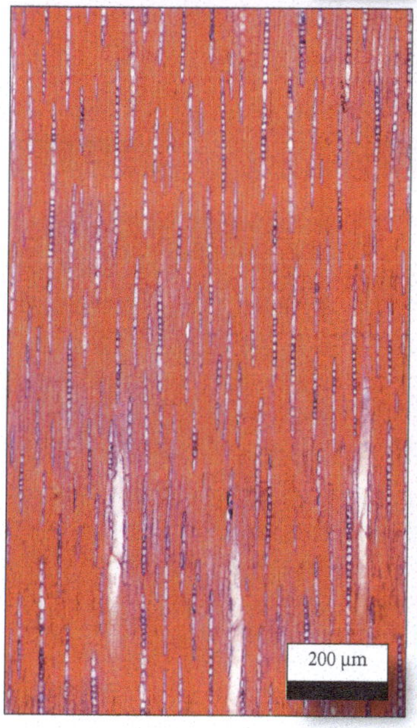

Tangential section

Rays exclusively uniseriate.
Ray height 130-397-850 µm.
Two to 4 cells per parenchyma strand.

Diospyros platycalyx Hiern

Synonymy: *D. striicalyx* H. Perrier

EN: Ebony   FR: Bois d'ébène   ML: Hazomainty

PLANT DESCRIPTION

Shrub to tree | 3-10 m high | 4-40 cm DBH

Endemic to Madagascar. Exploited species.

DISTRIBUTION

Elevation: 0-500 m a.s.l.

Vegetation formation: Forest and woodland

Bioclimate: Dry

Protected areas: Baie de Baly and Kirindy Menabe

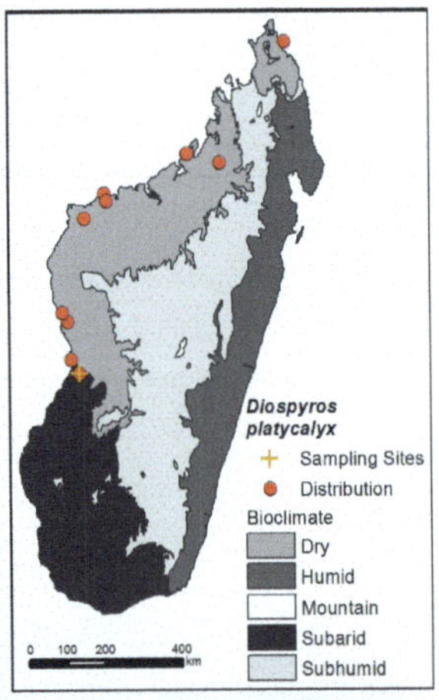

Slides made from material collected in Kirindy Menabe.

PITH AND PRIMARY XYLEM ANATOMY

Pith: Shape of pith round. A few scattered thick-walled parenchyma cells present. Cells dimorphic. Pits in transverse cell walls. Pith not visible in polarized light.

Primary xylem: Vascular bundles separated. Vessels solitary and in long radial multiples.

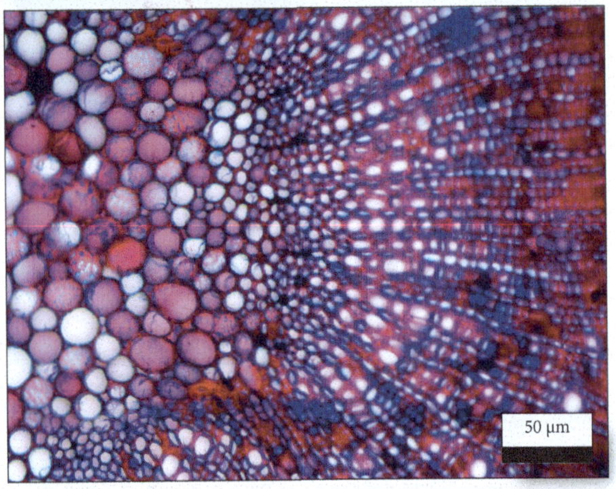

BRANCH XYLEM ANATOMY

Description: Growth rings boundaries indistinct or absent. Tangential diameter of vessels 20-38-50 µm, 112-160-204 vessels/mm^2. Vessel element length 220-299-440 µm. Dark staining substances present in heartwood vessel lumina. Intervessel pits small, 4-7 µm in diameter. Fibres thin- to thick-walled. Axial parenchyma diffuse-in-aggregates. Ray exclusively uniseriate. Ray height 310-417-560 µm. Rays 19-25-29/mm.

Anatomical differences to stem xylem: Mean tangential diameter of vessel lumina is similar in the branch and in the stem (20-38-50/25-48-60 µm). Vessel density is lower in the branch than in the stem (112-160-204/152-206-292 vessels/mm^2). Rays are higher in the branch (310-417-560/140-360-690 µm). Ray number is higher in the branch (19-25-29/16-19-23 rays/mm).

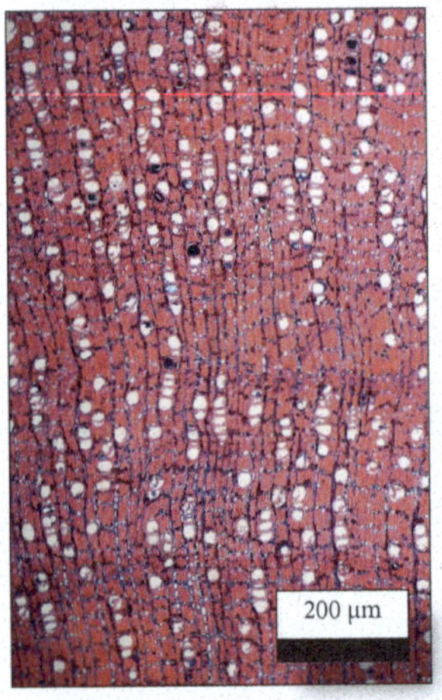

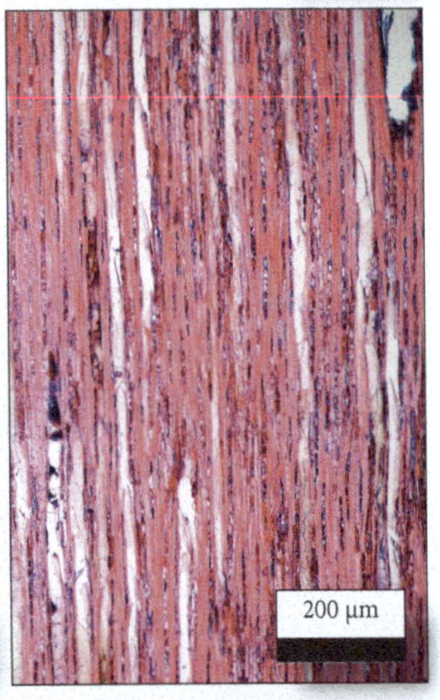

BARK ANATOMY

Branch bark: Collapsed and non-collapsed sieve tubes present in the phloem. Tangentially distributed solitary and groups of sclereids in the phloem. In the cortex sclereids arranged in a tangential bands.

Stem bark: Sieve tubes in tangential row. Collapsed and non-collapsed sieve tubes. Sclereids in groups, some are tangentially elongated. Prismatic crystals present in the sclereids. The phellem is homogeneous and it consists of regularly arranged rectangular cells, distinct in polarized light.

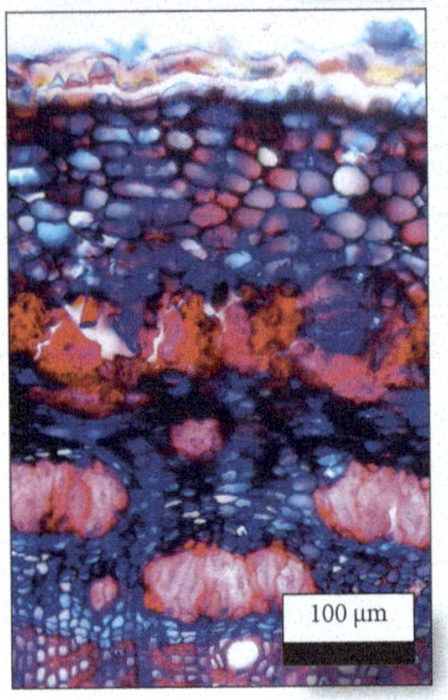

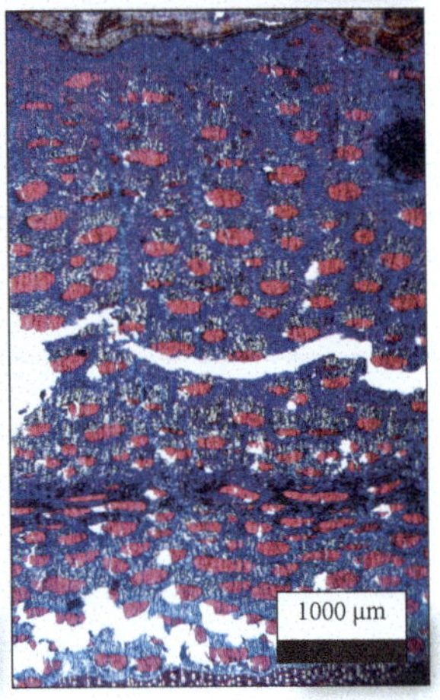

IAWA feature numbers
2 5 9.1 10 13 22 25 30 40 50 52 58 61 66 70 77 78 92
96.1 102.3 106 107 116 136 138 142 180 189 190 192 196

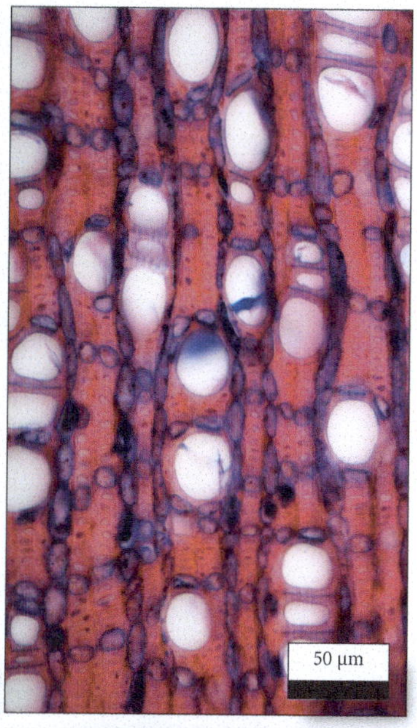

STEM XYLEM ANATOMY

Transverse section

Growth rings boundaries indistinct or absent.
Wood diffuse-porous.
Vessels solitary and in radial multiples of 2 to 6 elements.
Tangential diameter of vessels small 25-48-60 µm, 152-206-292 vessels/mm^2.
Dark staining substances present in heartwood vessel lumina.
Fibers very thick-walled.
Apotracheal parenchyma diffuse-in-aggregates, paratracheal parenchyma scanty.
Rays 16-19-23/mm.

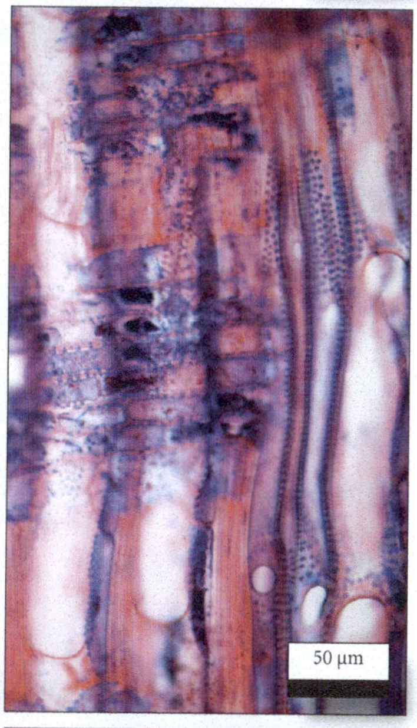

Radial section

Simple perforation plates.
Intervessel pits circular and alternate, 4-7 µm in diameter.
Vessel-ray pits with distinct borders; similar to intervessel pits in size and shape throughout the ray cell.
Vessel element length 200-276-360 µm.
Fibres with simple to minutely bordered pits (libriform fibers).
Heterogeneous rays, body ray cells procumbent with 1 to 4 rows of upright marginal cells.
Prismatic crystal in chambered axial parenchyma cells.

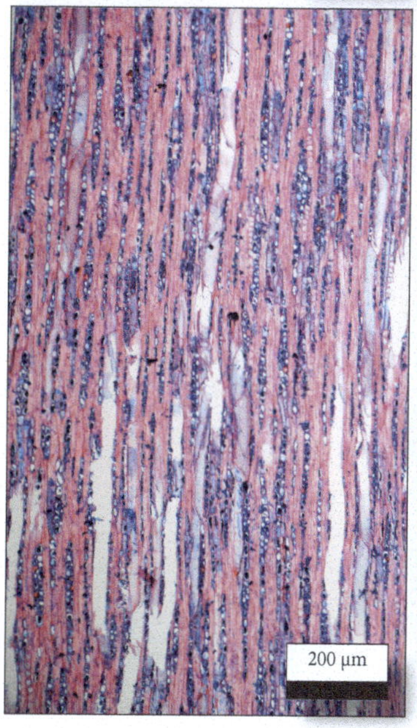

Tangential section

Rays predominantly uniseriate, a few biseriate rays present.
Ray height 140-360-690 µm.
Three to 4 cells per parenchyma strand.
Prismatic crystals in procumbent ray cells.

Diospyros quadrangularis ined.

EN: Ebony FR: Bois d'ébène ML: Hazomainty

PLANT DESCRIPTION

Tree | 7-20 m high | up to 27 cm DBH
Endemic to Madagascar. Exploited species.

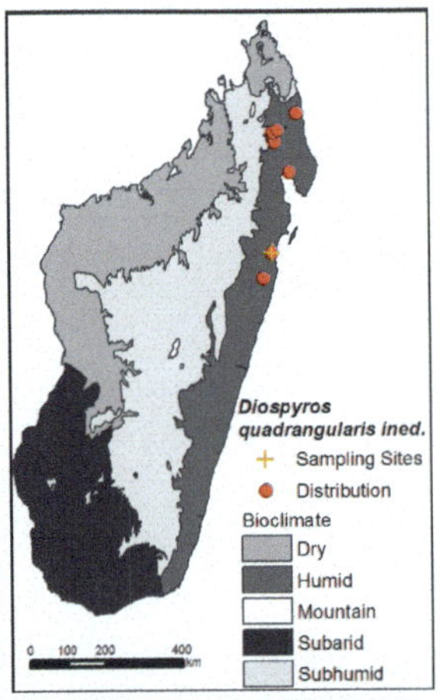

DISTRIBUTION

Elevation: 0-1500 m a.s.l.
Vegetation formation: Forest
Bioclimate: Humid
Protected areas: Betampona
New protected areas: Makirovana and Tampolo

Slides made from material collected in Tampolo.

PITH AND PRIMARY XYLEM ANATOMY

Pith: Shape of pith oval. Cells dimorphic. Dark staining substances fill the lumen of some cells. Prismatic crystals present. Pits in transverse cell walls. Pith not visible in polarized light.

Primary xylem: Vascular bundles not distinct. Vessels solitary and in short radial multiples.

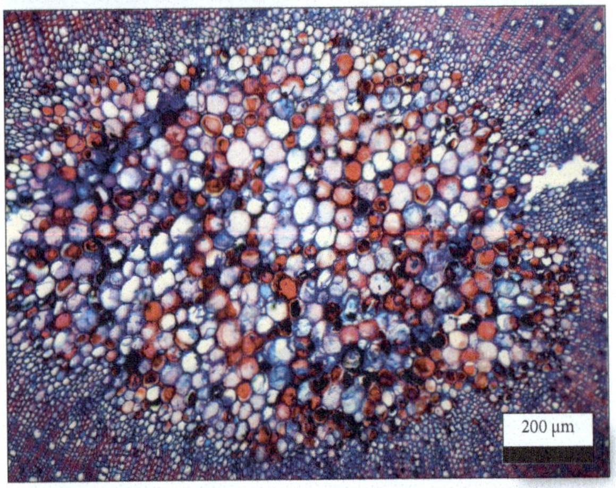

BRANCH XYLEM ANATOMY

Description: Growth ring boundaries absent to slightly visible. Vessels predominantly solitary. Tangential diameter of vessels 20-35-50 µm, 20-47-88 vessels/mm². Vessel element length 280-320-420 µm. Intervessel pits small, 4-7 µm in diameter. Fibres thin- to thick-walled. Apotracheal parenchyma diffuse-in-aggregates, paratracheal parenchyma scanty to vasicentric. Rays exclusively uniseriate. Rays with procumbent, squared and upright cells mixed throughout the ray. Ray height 270-511-750(1300) µm. Rays 16-20-25/mm. Prismatic crystals in axial parenchyma cells.

Anatomical differences to stem xylem: Mean tangential diameter of vessel lumina is smaller in the branch than in the stem (20-35-50/50-72-90 µm). Vessel density and ray height are higher in the branch than in the stem (20-47-88/16-26-40 vessels/mm²; 270-511-750/200-400-980 µm). Ray number is smaller in the branch than in the stem (16-20-25/16-26-40 rays/mm).

BARK ANATOMY

Branch bark: Sieve tubes in tangential rows. Some rays get dilated. Small groups of thick-walled fibers in the cortex. Sclereids in groups in the phloem and in the cortex. Prismatic crystals in tangential lines in phloem, irregularly dispersed in the cortex. Dark staining substances fills phellem cells.

Stem bark: The phloem is composed by collapsed and non-collapsed sieve tubes. Sclereids in groups, tangentially arranged towards the outside. The phellem is homogeneous, dark stained substances fills phellem cell lumina. Phellem indistinct in polarized light. Prismatic crystals in sclereids.

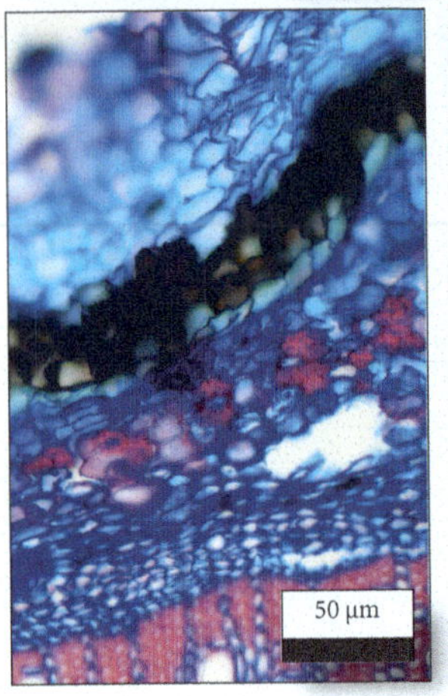

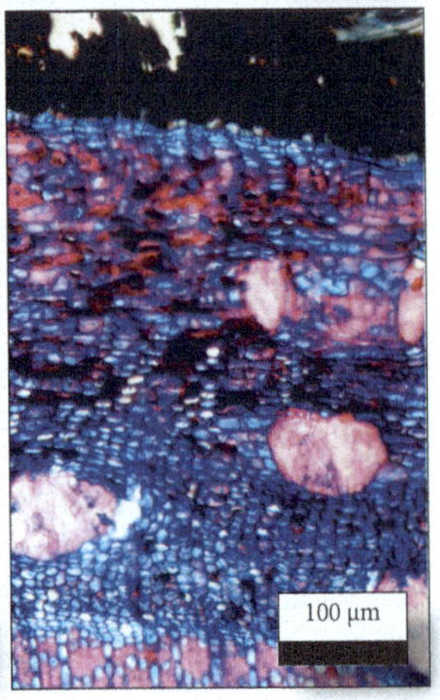

IAWA feature numbers

2 5 9 10 13 22 25 30 41 48 53 61 66 70 78 79 86 92 96
102.3 107 108 116 136 142 180 189 192 196

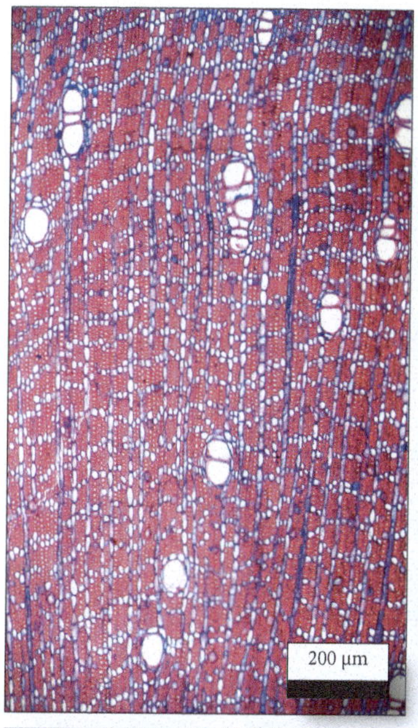

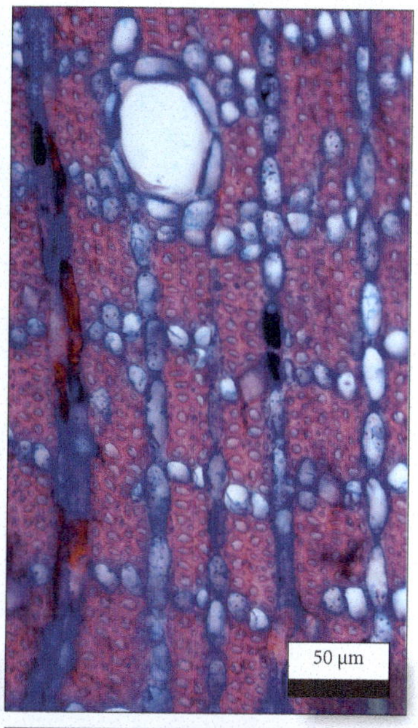

STEM XYLEM ANATOMY

Transverse section

Growth rings boundaries indistinct or absent.
Wood diffuse-porous.
Vessels solitary and in radial multiples of 2-6 elements. Tangential diameter of vessels 50-72-90 µm, 16-26-40 vessels/mm^2.
Fibers very thick-walled.
Apotracheal parenchyma in narrow lines up to 3 cells, paratracheal parenchyma scanty to vasicentric.
Rays 17-20-24/mm.

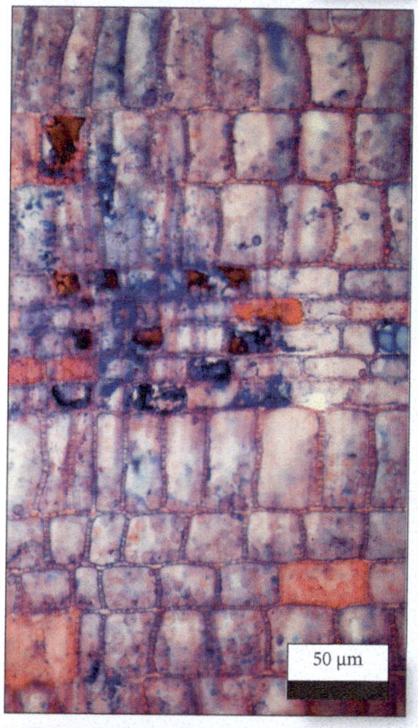

Radial section

Simple perforation plates.
Intervessel pits circular and alternate, 4-7 µm in diameter.
Vessel-ray pits with distinct borders; similar to intervessel pits in size and shape throughout the ray cell. Vessel element length 220-414-600 µm.
Fibres with simple to minutely bordered pits (libriform fibers).
Heterogeneous rays, body ray cells procumbent with over 4 rows of upright marginal cells.
Prismatic crystals in chambered axial parenchyma cells.

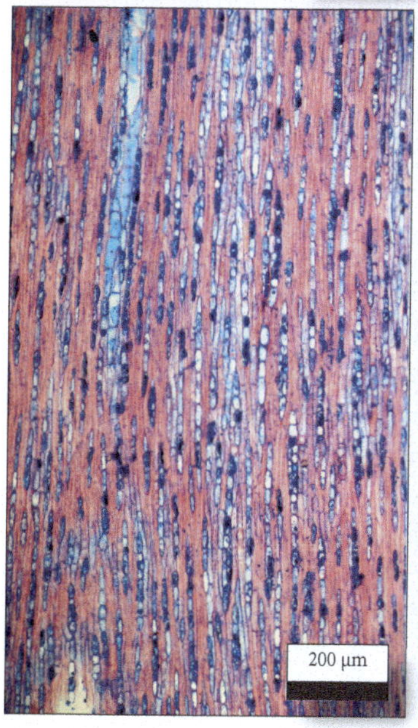

Tangential section

Rays exclusively uniseriate.
Ray height 200-400-980 µm.
Two to 4 cells per parenchyma strand.

Diospyros rubripetiolata ined.

EN: Ebony FR: Bois d'ébène ML: Hazomainty

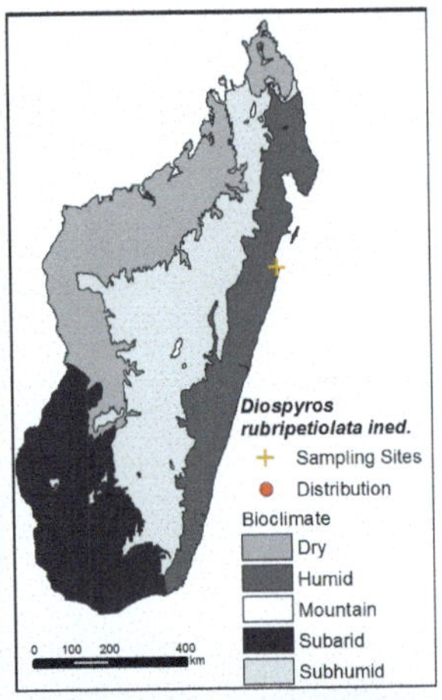

PLANT DESCRIPTION

Shrub to tree | 7-15 m high | up to 18 cm DBH
Endemic to Madagascar. Non-exploited species.

DISTRIBUTION

Elevation: 0-500 m a.s.l.
Vegetation formation: Forest
Bioclimate: Humid
Protected areas: -
New protected areas: Andranotsara

Slides made from material collected in Andranotsara.

PITH AND PRIMARY XYLEM ANATOMY

Pith: Shape of pith round. Thick-walled parenchyma cells present. Cells dimorphic. Prismatic crystals present. Pits in transverse cell walls. Pith visible in polarized light.

Primary xylem: Vascular bundles not distinct. Vessels solitary and in short radial multiples of 2 to 4 elements.

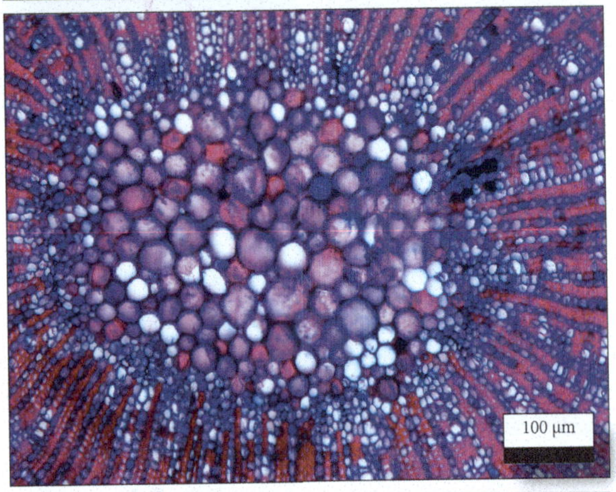

BRANCH XYLEM ANATOMY

Description: Growth ring boundaries absent to slightly visible. Vessels mostly in short radial multiples of 2 to 4(6) elements. Tangential diameter of vessels 30-43-60 µm, 48-59-67 vessels/mm². Vessel element length 200-376-490 µm. Intervessel pits small, 4-7 µm in diameter. Fibres thin- to thick-walled. Apotracheal parenchyma diffuse-in-aggregates, paratracheal parenchyma vasicentric. Rays mostly uniseriate. Ray height 260-539-1000 µm. Rays 13-15-17/mm. Prismatic crystals in axial parenchyma cel ls.

Anatomical differences to stem xylem: Mean tangential diameter of vessel lumina is smaller in the branch (30-43-60/40-59-80 µm). Vessel density in higher in the branch (17-25-34/48-59-67 vessels/mm²). The number of rays is slightly higher in the branch (13-15-17/11-14-16 rays/mm). Rays are slightly shorter in branch (260-539-1000/300-658-1020 µm).

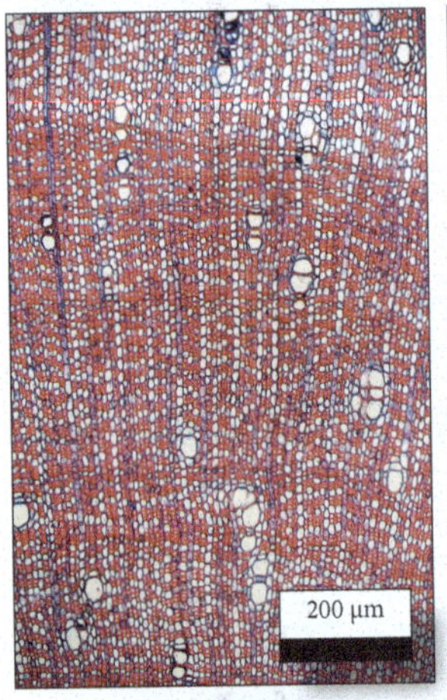

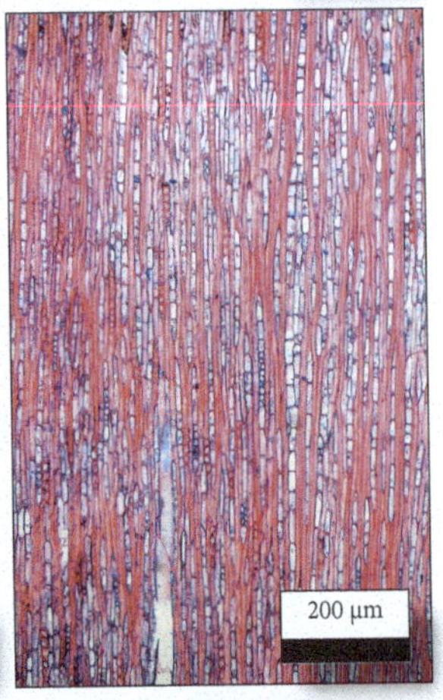

BARK ANATOMY

Branch bark: Collapsed and non-collapsed sieve tubes present in the phloem. The cortex is characterized by a band of sclereids and irregularly shaped blue- and red-stained parenchyma cells.

Stem bark: Sieve tubes grouped. A few rays become slightly dilated. Sclereids are scattered and grouped in irregular groups. Prismatic crystals present in sclereids and in axial parenchyma cells. The phellem is homogeneous with dark stained substances and it consists of regularly arranged rectangular cells, indistinct in polarized light.

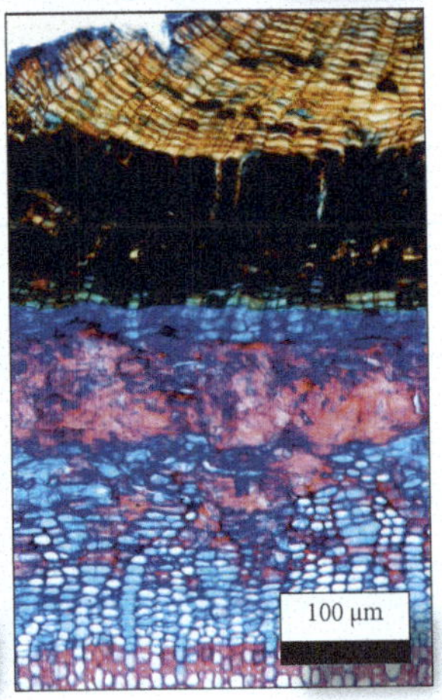

IAWA feature numbers

1 5 9 13 22 25 30 41 48 52 61 66 69 77 79 91 92 97 100
102.4 108 116 136 142 154 180 189 192 196 202

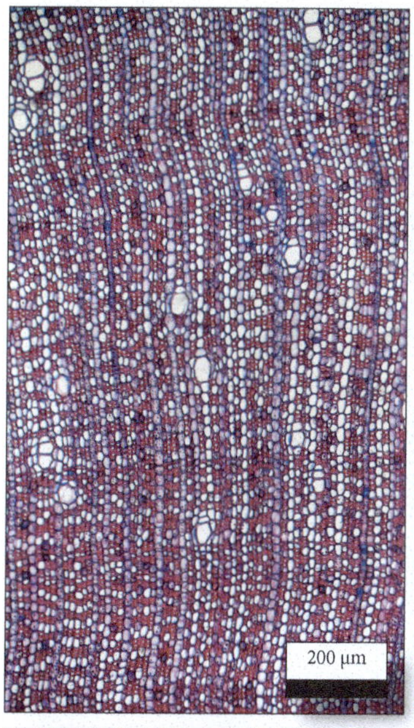

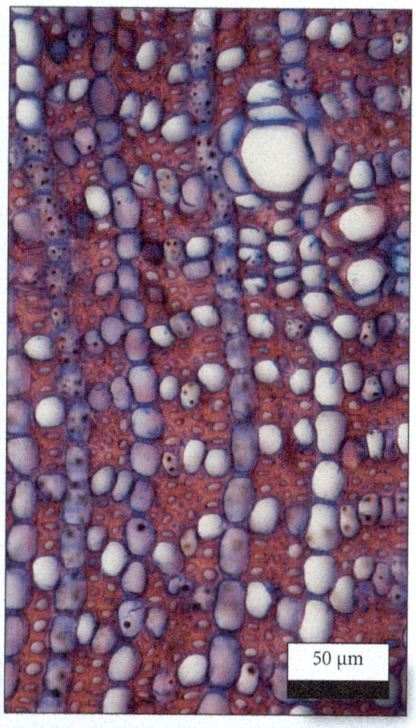

STEM XYLEM ANATOMY

Transverse section

Growth ring boundaries distinct.
Wood diffuse-porous.
Vessels predominantly solitary.
Tangential diameter of vessels 40-59-80 µm, 17-25-34 vessels/mm^2.
Fibers thin- to thick-walled.
Axial parenchyma diffuse-in-aggregates and vasicentric.
Rays 11-14-16/mm.

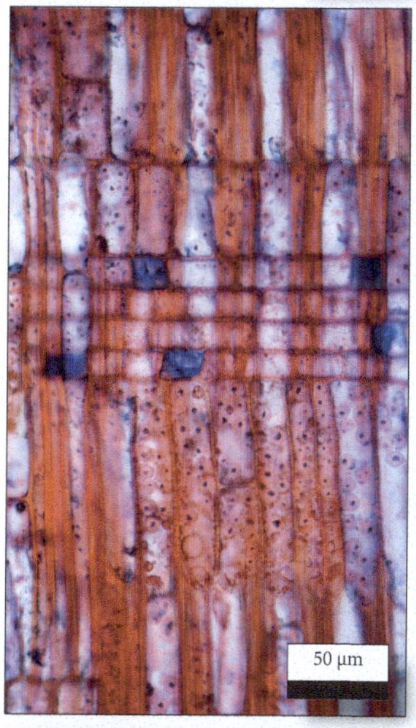

Radial section

Simple perforation plates.
Intervessel pits circular and alternate, 4-7 µm in diameter. Vessel-ray pits with distinct borders; similar to intervessel pits in size and shape throughout the ray cell.
Vessel element length 190-244-350 µm.
Fibres with simple to minutely bordered pits (libriform fibers).
Body ray cells procumbent with 2 to over 4 rows of upright and square marginal cells.
Prismatic crystal in chambered axial parenchyma cells. Crystals frequent.

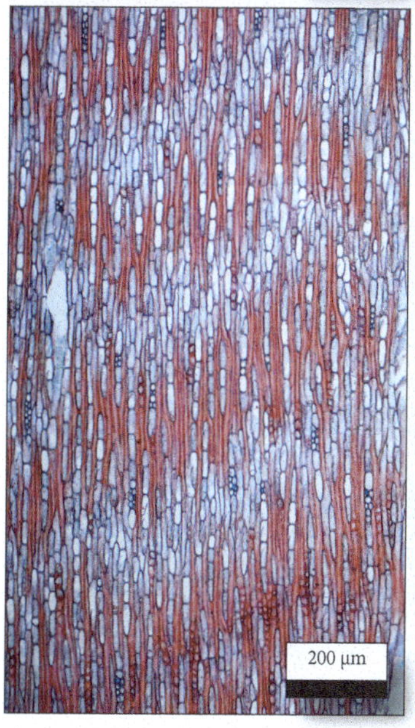

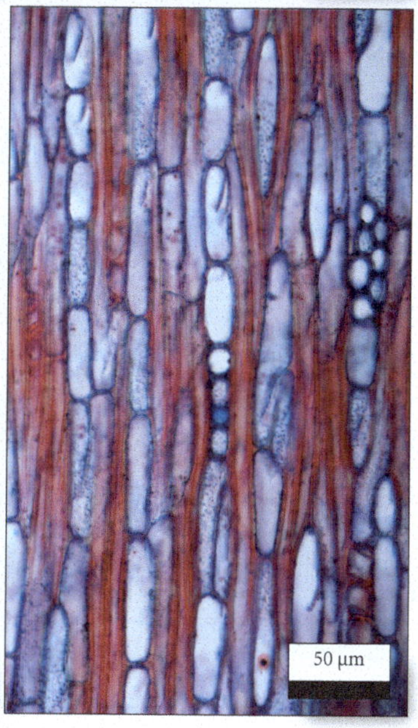

Tangential section

Rays uni- and biseriate, predominantly biseriate.
Ray height 300-658-1020 µm.
Rays with multiseriate portion as wide as uniseriate portion.
Two to 4 cells per parenchyma strand.

Diospyros sclerophylla H. Perrier

EN: Ebony FR: Bois d'ébène ML: Hazomainty

PLANT DESCRIPTION

Tree | 5-12 m high | up to 30 cm DBH
Endemic to Madagascar. Exploited species.

DISTRIBUTION

Elevation: 0-500 m a.s.l.
Vegetation formation: Forest
Bioclimate: Humid
Protected areas: Ankarafantsika
New protected areas: Tampolo

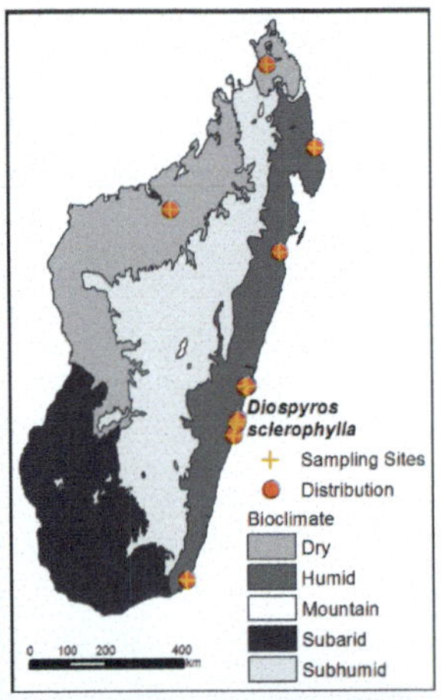

Slides made from material collected in Tampolo.

PITH AND PRIMARY XYLEM ANATOMY

Pith: Shape of pith round to oval. Thick-walled parenchyma cells present. Cells dimorphic. Prismatic crystals present. Pits in transverse cell walls. Pith visible in polarized light.

Primary xylem: Vascular bundles not distinct. Vessels solitary and in long radial multiples.

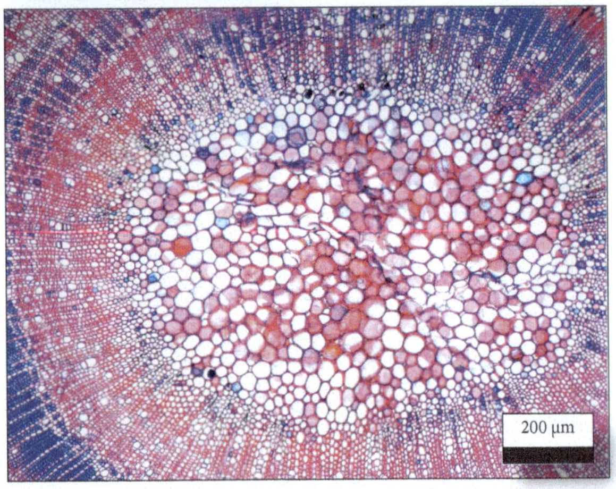

BRANCH XYLEM ANATOMY

Description: Growth rings boundaries indistinct or absent. Vessels predominantly solitary. Tangential diameter of vessels 40-65-80 µm, 24-30-34 vessels/mm^2. Vessel element length 160-244-440 µm. Intervessel pits small, 4-7 µm in diameter. Fibres thin- to thick-walled. Axial parenchyma diffuse-in-aggregates, and scanty paratracheal. Ray predominantly uniseriate. Ray height 120-420-750 µm. Rays 14-17-22/mm. Rare crystals in axial parenchyma cells.

Anatomical differences to stem xylem: Mean tangential diameter of vessel lumina is smaller in the branch (40-65-80/60-84-110 µm), vessel density is smaller in the branch (24-30-34/22-33-38 vessels/mm^2). Rays are shorter in the branch (120-420-750/270-538-1120 µm), whereas they are more numerous (14-17-22/8-11-15 rays/mm).

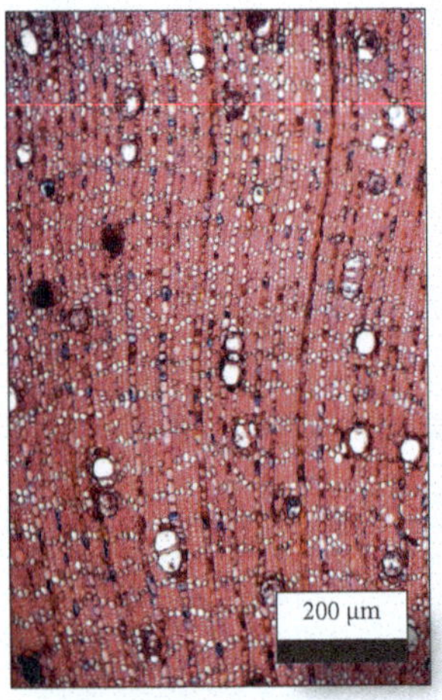

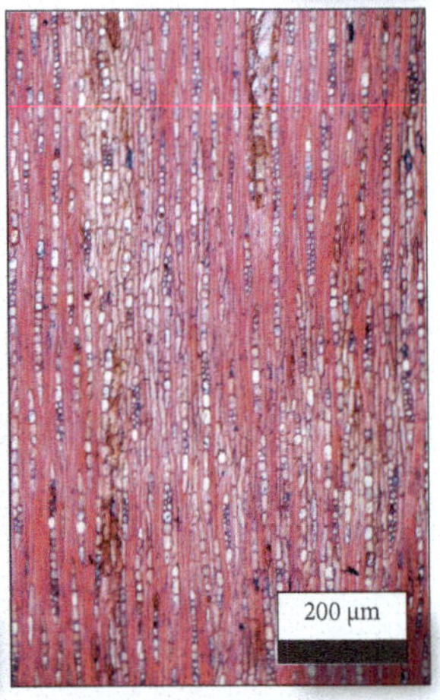

BARK ANATOMY

Branch bark: Collapsed and non-collapsed sieve tubes present in the phloem. The cortex is characterized by a band of sclerenchyma and irregular blue-stained parenchyma cells.

Stem bark: Sieve tubes in tangential rows, collapsed sieve tubes present. Distinct ray dilation occur in the outermost phloem. A few large groups of sclereids present. Prismatic crystals present in the sclereids and in ray parenchyma cells. The phellem is homogeneous and consists of regularly arranged rectangular cells, indistinct in polarized light. Dark staining substances in some phellem cells.

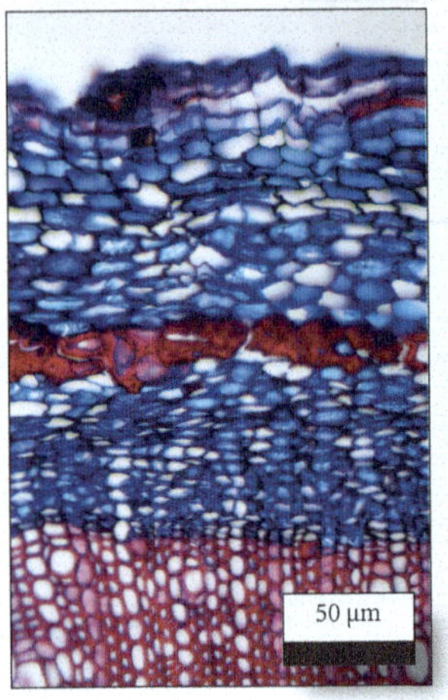

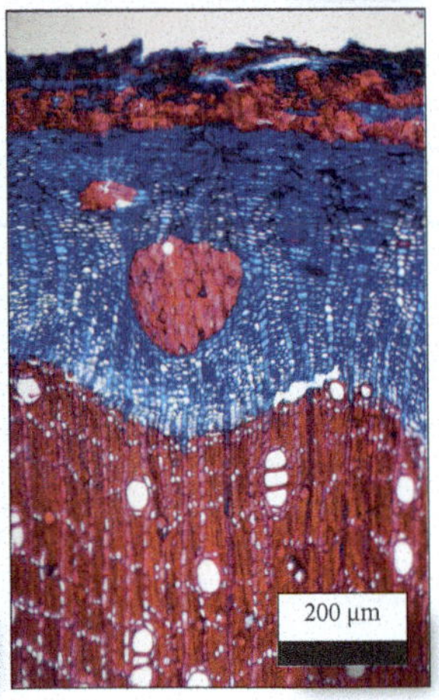

IAWA feature numbers
2 5 9 9.1 13 22 25 30 41 47 52 61 66 69 77 78 86 92 96.1
102.4 107 108 115 136 142 154 180 189 192 196 202

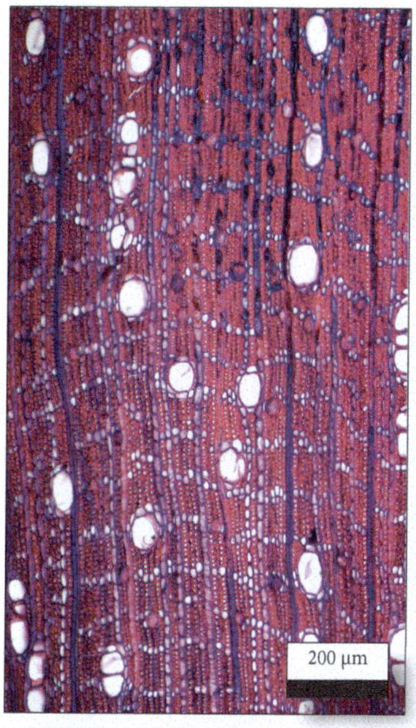

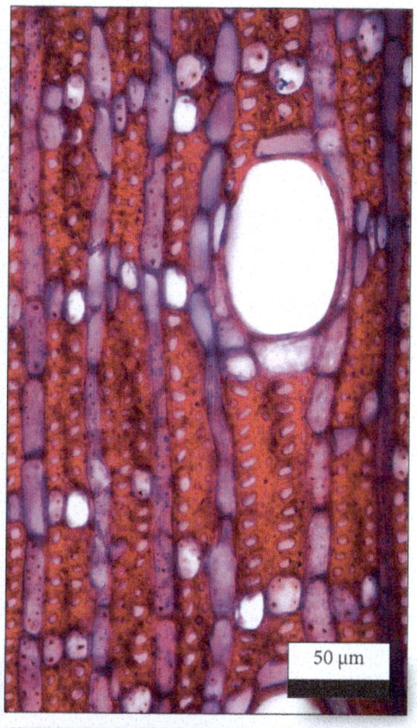

STEM XYLEM ANATOMY

Transverse section

Growth rings boundaries indistinct or absent.
Wood diffuse-porous.
Vessels solitary and in short radial multiples of 2-3 elements.
Tangential diameter of vessels 60-84-110 µm, 25-33-38 vessels/mm^2.
Fibers thick-walled.
Apotracheal parenchyma diffuse-in-aggregates.
Paratracheal parenchyma scanty.
Rays 18-22-25/mm.

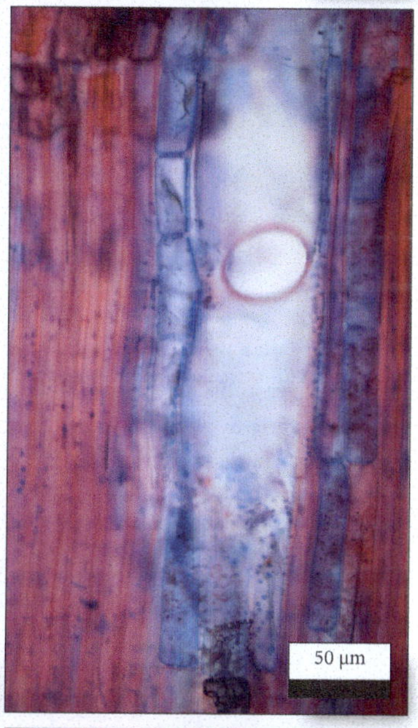

Radial section

Simple perforation plates.
Intervessel pits circular and alternate, 4-7 µm in diameter.
Vessel-ray pits with distinct borders; similar to intervessel pits in size and shape throughout the ray cell.
Vessel element length 230-350-560 µm.
Fibres with simple to minutely bordered pits (libriform fibers).
Heterogeneous rays with body ray cells procumbent and 2 to over 4 rows of upright and square marginal cells.
One prismatic crystal per chamber in axial parenchyma cells. More than one crystal per cell. Crystals rare.

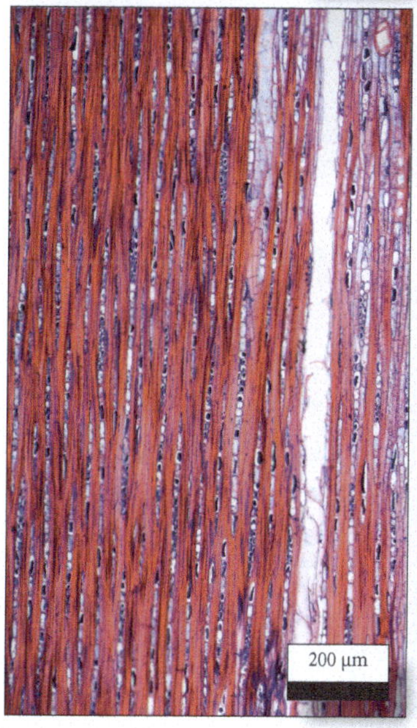

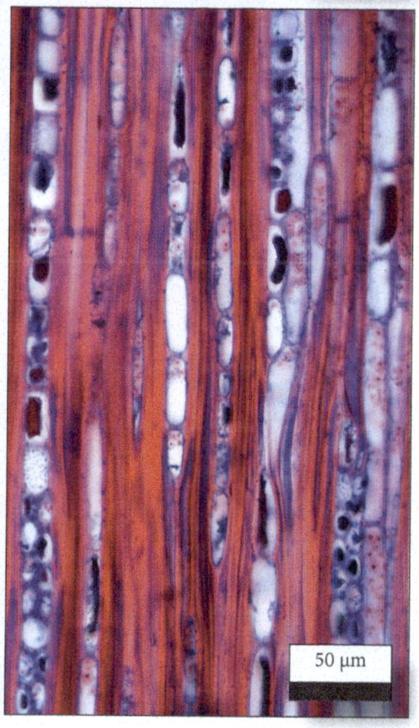

Tangential section

Rays predominantly uniseriate, a few biseriate rays present.
Ray height 270-538-1120 µm.
Two to 4 cells per parenchyma strand.

Diospyros squamosa Bojer ex A. DC.

Synonymy: *D. bernieri* Hiern; *D. laevis* Bojer ex A. DC.

EN: Ebony FR: Bois d'ébène ML: Hazomainty

PLANT DESCRIPTION

Tree | up to 20 m high | 15-28 cm DBH
Endemic to Madagascar. Exploited species.

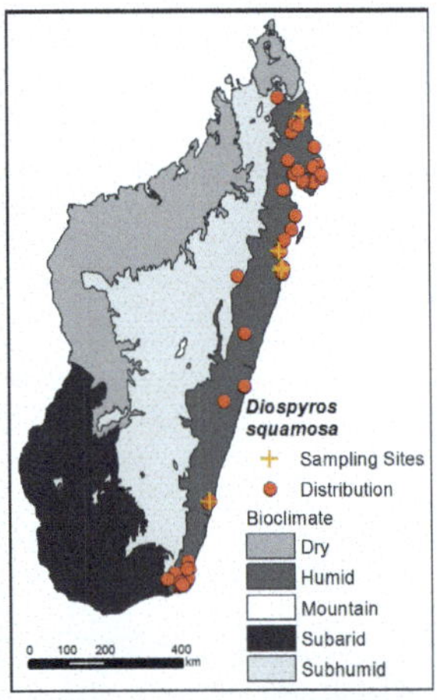

DISTRIBUTION

Elevation: 0-500 m a.s.l.
Vegetation formation: Forest
Bioclimate: Humid
Protected areas: Andohahela, Manombo, Marojejy, Masoala and Nosy Mangabe
New protected areas: Tampolo, Analalava, Andranotsara, Mahabo-Mananivo and Makirovana

Slides made from material collected in Andranotsara.

PITH AND PRIMARY XYLEM ANATOMY

Pith: Shape of pith oval to polygonal. Thick-walled parenchyma cells present. Cells dimorphic. Prismatic crystals present. Pits in transverse cell walls. Pith visible in polarized light.

Primary xylem: Vascular bundles not distinct. Vessels solitary and in long radial multiples of 3 to 10 elements.

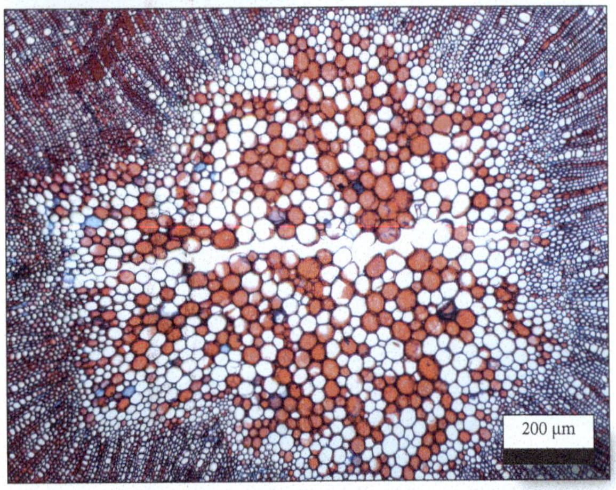

BRANCH XYLEM ANATOMY

Description: Growth rings boundaries indistinct or absent. Vessels mainly organized in short radial multiples. Tangential diameter of vessels 40-57-80 µm, 27-36-50 vessels/mm^2. Vessel element length 150-246-320 µm. Intervessel pits small, 4-7 µm in diameter. Fibres thin- to thick-walled. Apotracheal parenchyma in narrow bands up to 3 cells wide. Paratracheal parenchyma scanty. Rays exclusively uniseriate, ray height 130-259-460 µm. Rays 16-18-22/mm. Prismatic crystals in chambered axial parenchyma cells. More than one crystal per cell.

Anatomical differences to stem xylem: Mean tangential diameter of vessel lumina is slightly smaller in the branch (40-57-80/80-96-120 µm), while vessel density is slightly higher in the branch (27-36-50/24-31-40 vessels/mm^2). Ray height is smaller in the branch (130-259-460/140-373-920 µm), whereas the number of rays is similar (16-18-22/15-18-22 rays/mm).

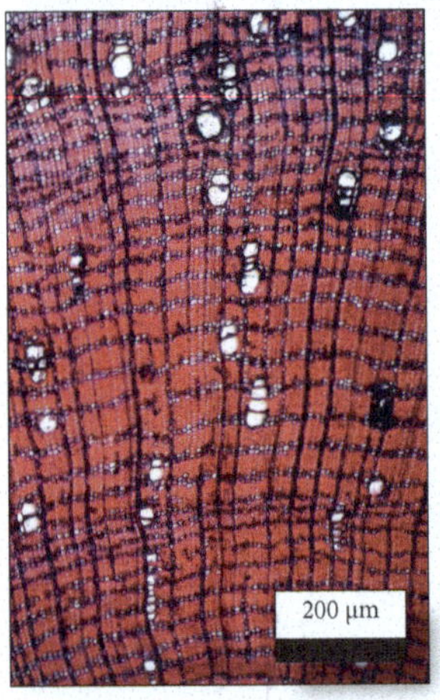

BARK ANATOMY

Branch bark: Collapsed and non-collapsed sieve tubes present in the phloem. Some rays become dilated. The cortex is characterized by a band of sclereids and irregular blue-stained parenchyma cells. Dark staining substances in phellem cell lumina.

Stem bark: The phloem is characterized by sieve tubes and parenchyma cells filled with dark staining substances. Sclereids in irregularly distributed groups. Prismatic crystals present within sclereids groups. Crystals are not frequent. The phellem cells are homogeneous and it consists of regularly arranged rectangular cells, indistinct in polarized light. Dark staining substances in phellem cells.

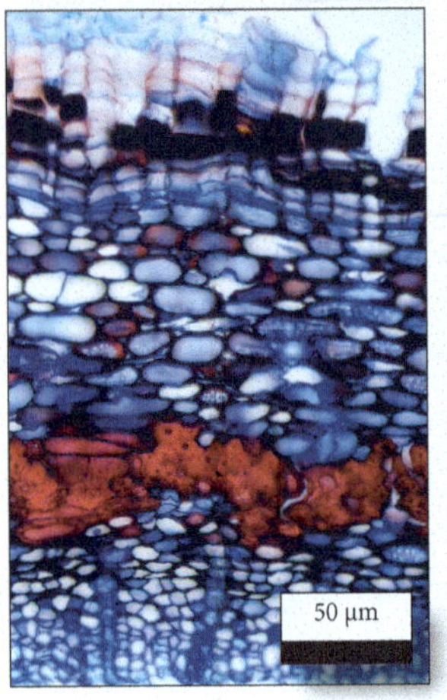

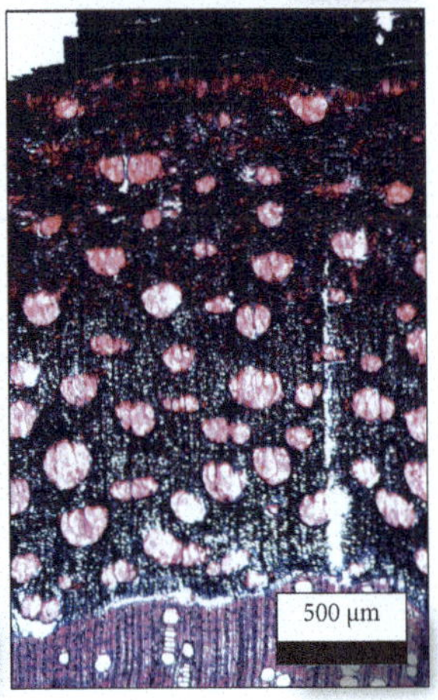

IAWA feature numbers

2 5 9 9.1 13 22 25 30 41 42 48 52 61 66 69 78 86 91 92 96.1 102.3 107 116 136 142 154 180 189 192 196 202

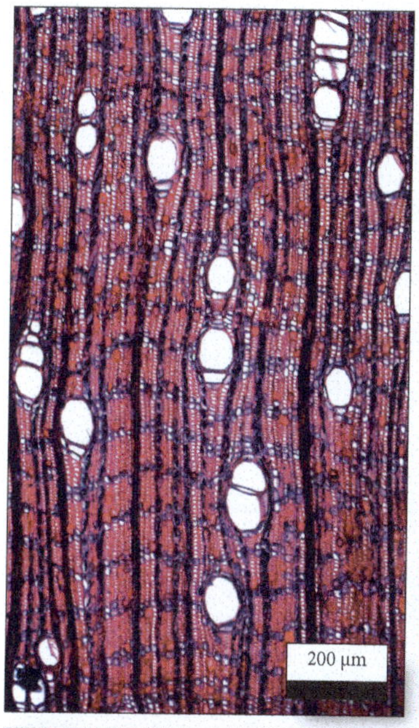

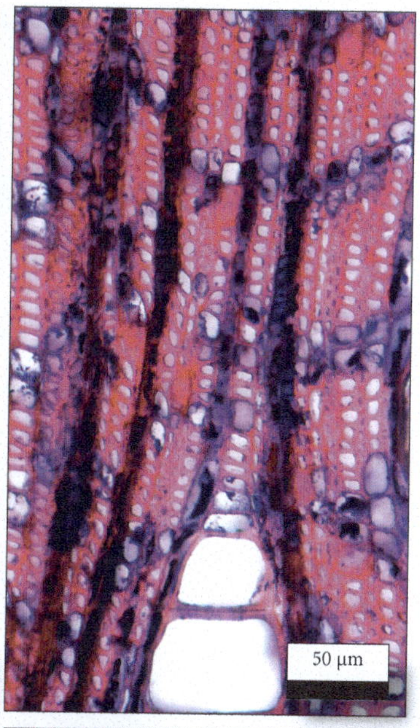

STEM XYLEM ANATOMY

Transverse section

Growth ring boundaries absent.
Wood diffuse-porous.
Vessels solitary and in radial multiples.
Tangential diameter of vessels medium 80-96-120 µm, 24-31-40 vessels/mm^2.
Fibers thin- to thick-walled.
Apotracheal parenchyma diffuse-in-aggregates, and in narrow lines up to 3 cells.
Paratracheal parenchyma scanty.
Rays 15-18-22/mm.

Radial section

Simple perforation plates.
Intervessel pits circular and alternate, 4-7 µm in diameter.
Vessel-ray pits with distinct borders; similar to intervessel pits in size and shape throughout the ray cell.
Vessel element length 160-269-480 µm.
Fibres with simple to minutely bordered pits (libriform fibers).
Heterogeneous rays with body ray cells procumbent and 2 to over 4 rows of upright and square marginal cells.
Prismatic crystal in chambered axial parenchyma cells. More than one crystal per cell. Crystals frequent.

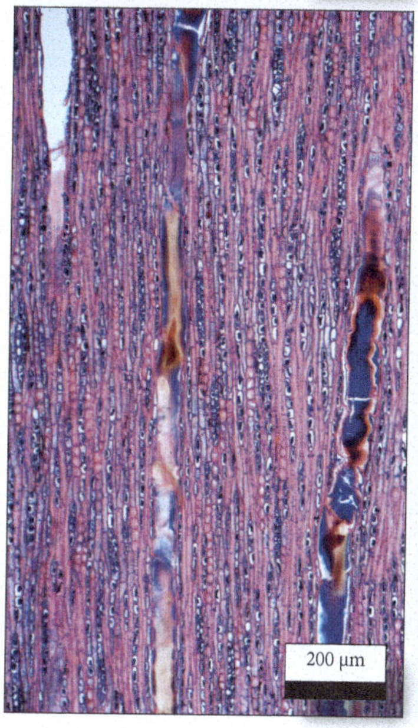

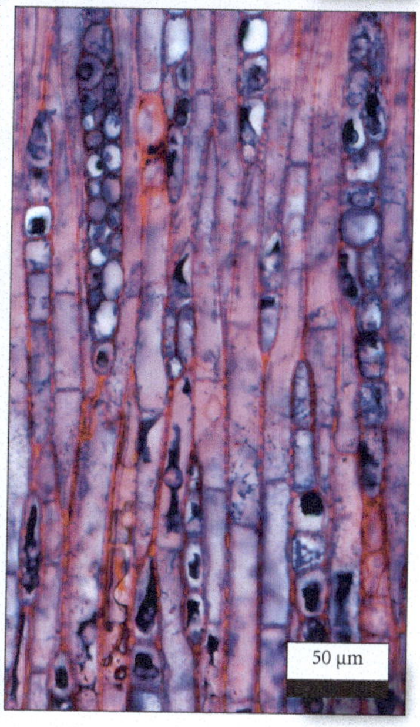

Tangential section

Rays predominantly biseriate, a few uniseriate rays present.
Ray height 140-373-920 µm.
Two to 4 cells per parenchyma strand.

Diospyros tropophylloides ined.

EN: Ebony FR: Bois d'ébène ML: Hazomainty

PLANT DESCRIPTION

Shrub to tree | up to 2-14 m high | up to 15 cm DBH
Endemic to Madagascar. Non-exploited species.

DISTRIBUTION

Elevation: 0-500 m a.s.l.
Vegetation formation: Forest
Bioclimate: Humid, Dry
Protected areas: Ankarana
New protected areas: Makirovana, Orangea/Rigny and Montagne des Français

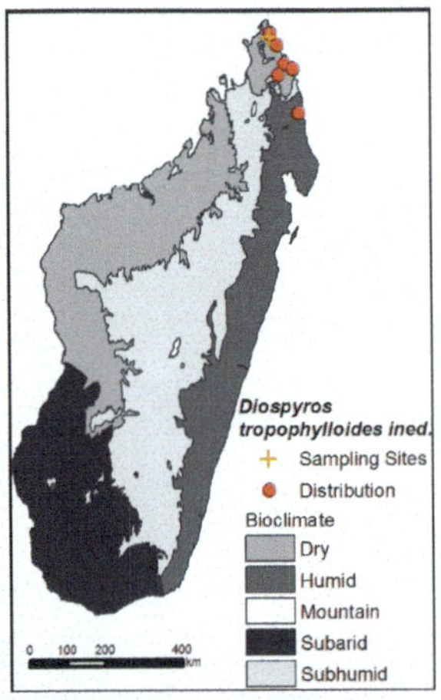

Slides made from material collected in Montagne des Français.

PITH AND PRIMARY XYLEM ANATOMY

Pith: Shape of pith round. Thick-walled parenchyma cells present, predominating in number. Cells dimorphic. Cell content in a few cells. Prismatic crystals present. Pits in transverse cell walls. Pith visible in polarized light.

Primary xylem: Vascular bundles not clearly distinct. Vessels solitary and in short radial multiples of 2 to 4 elements.

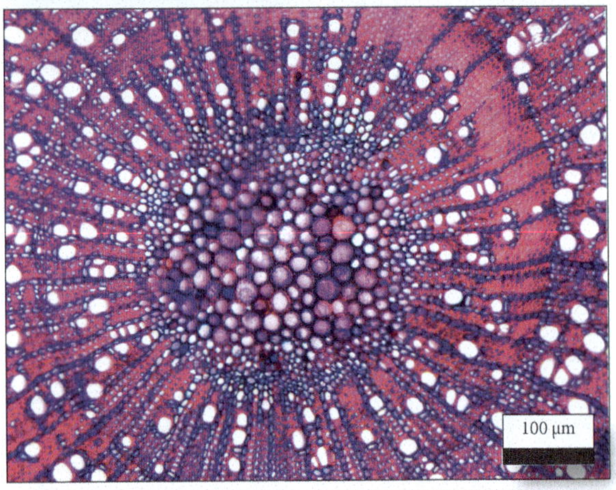

BRANCH XYLEM ANATOMY

Description: Growth rings boundaries indistinct or absent. Vessels solitary and in radial multiples of 2 to 8(12) elements. Tangential diameter of vessels 20-34-40 µm, 160-196-240 vessels/mm^2. Intervessel pits small, 4-7 µm in diameter. Vessel element length 160-288-430 µm. Gums and other deposits in vessels. Fibres thin- to thick-walled. Apotracheal parenchyma diffuse-in-aggregates, paratracheal parenchyma scanty. Rays exclusively uniseriate, ray height 80-351-720 µm. Body ray cells procumbent with one row of upright marginal cells. Rays 13-16-21/mm. Prismatic crystals in axial parenchyma cells.

Anatomical differences to stem xylem: Mean tangential diameter of vessel lumina is similar in branch and in stem xylem (20-34-40/20-35-50 µm), while vessel density is slightly higher in the stem (160-196-240/230-292-348 vessels/mm^2). Rays are similar in height (80-351-720/250-350-500 µm), while ray number is higher in branch xylem (11-16-21/10-13-17 rays/mm).

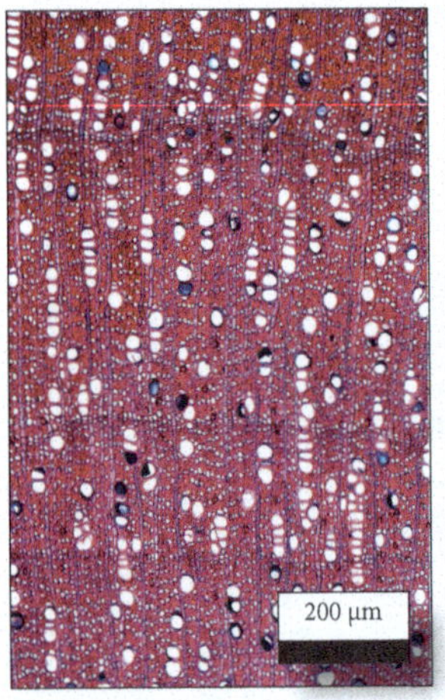

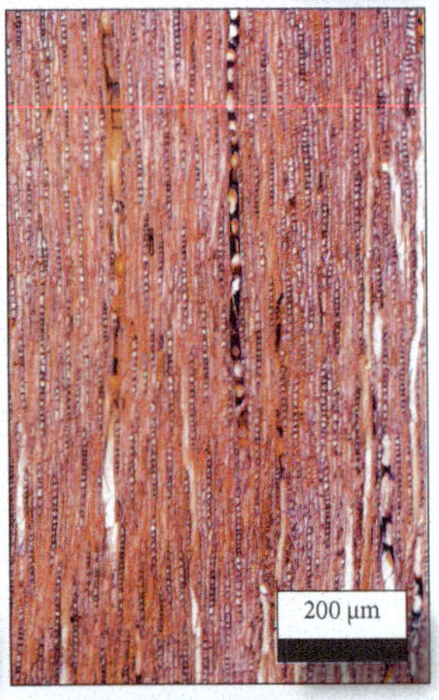

BARK ANATOMY

Branch bark: Collapsed and non-collapsed sieve tubes present in the phloem. Only some rays become dilated. The cortex is characterized by sclereids in tangential bands and irregular blue-stained parenchyma cells.

Stem bark: Sieve tubes in tangential rows alternating with axial parenchyma cells. Ray dilatations present. Sclereids in tangentially elongated and regularly distributed groups. Many prismatic crystals present in the sclereids and scattered or irregularly dispersed in the parenchyma cells. The phellem is layered and homogeneous, it consists of regularly arranged rectangular cells, indistinct in polarized light. Cell content in phellem cells.

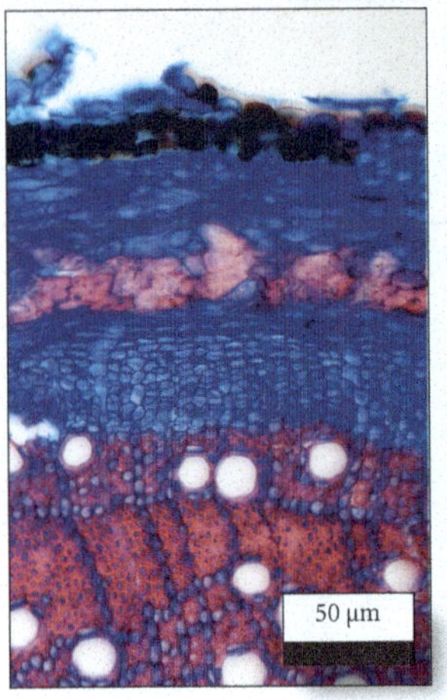

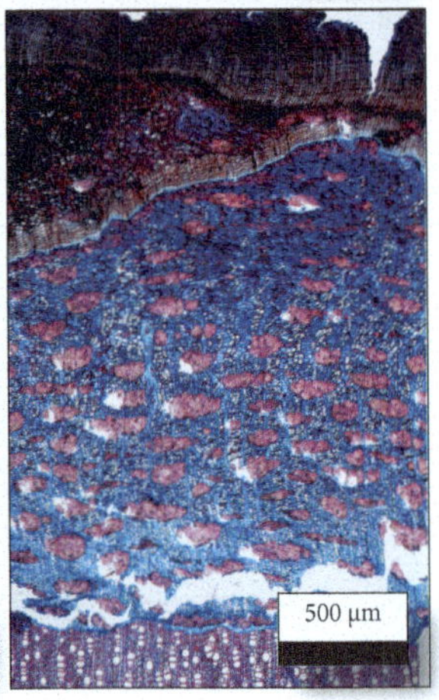

IAWA feature numbers

2 5 7 9.1 10 13 22 25 30 40 50 52 58 61 66 69 70 77 78 92 96.1 102.3 106 107 116 136 142 152 180 189 190 192 196

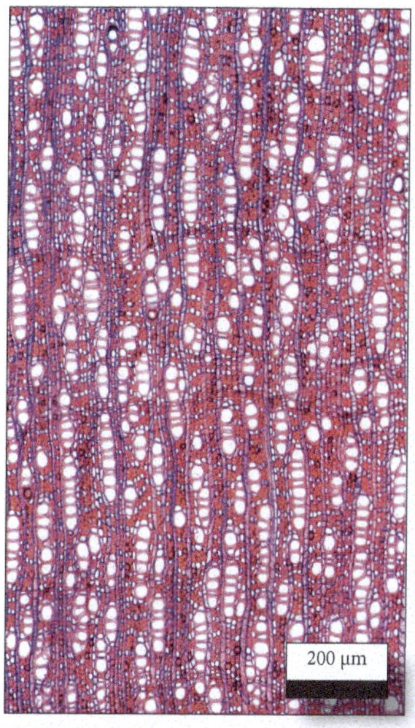

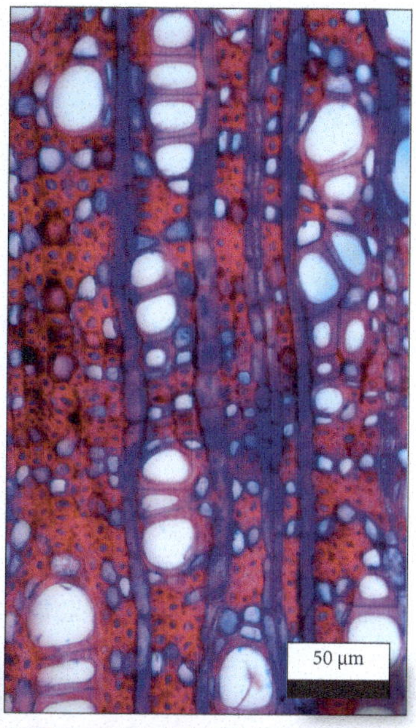

STEM XYLEM ANATOMY

Transverse section

Growth rings boundaries indistinct or absent.
Wood diffuse-porous.
Vessels in radial pattern, mainly in radial multiples of 2 to 8(12) elements.
Tangential diameter of vessels small 30-37-50 µm, 230-292-348 vessels/mm^2.
Gums present in heartwood vessel lumina.
Fibers thick- to very thick-walled.
Apotracheal parenchyma diffuse-in-aggregates, paratracheal parenchyma scanty.
Rays 10-13-17/mm.

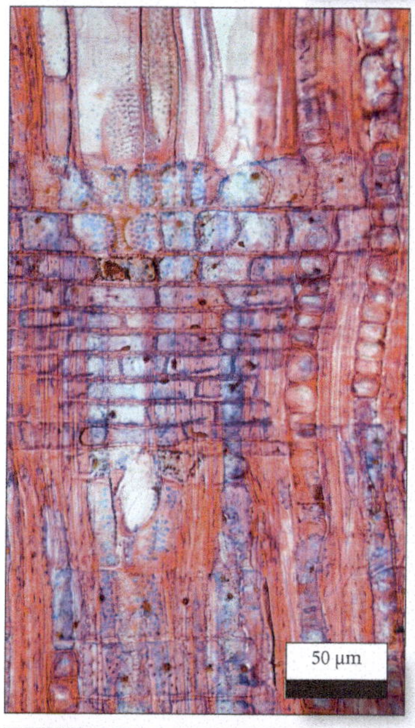

Radial section

Simple perforation plates.
Intervessel pits circular and alternate, 4-7 µm in diameter.
Vessel-ray pits with distinct borders; similar to intervessel pits in size and shape throughout the ray cell.
Vessel element length 220-322-380 µm.
Fibres with simple to minutely bordered pits (libriform fibers).
Heterogeneous rays with body ray cells procumbent and 1 to 4 rows of upright marginal cells.
Prismatic crystals in chamber axial parenchyma cells. More than one crystal per cell. Crystals frequent.

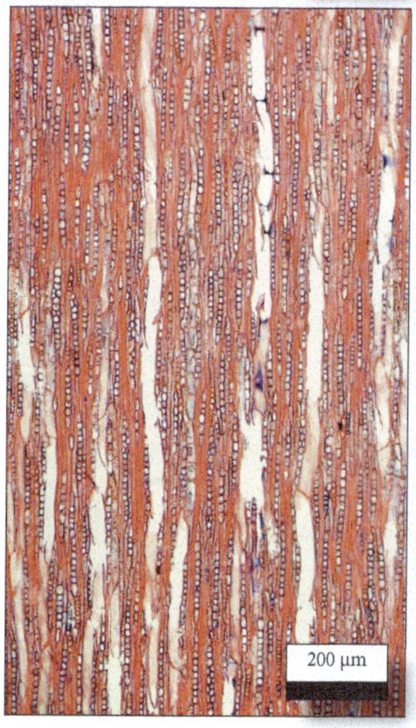

Tangential section

Rays predominantly uniseriate, a few biseriate rays present.
Ray height 250-350-500 µm.
Tree to 4 cells per parenchyma strand.

Diospyros velutipes (H. Perrier) G. E. Schatz and Lowry

Synonymy: *D. gracilipes* var. *velutipes* H. Perrier

EN: Ebony FR: Bois d'ébène ML: Hazomainty

PLANT DESCRIPTION

Tree | up to 5-12 m high | 5-10 cm DBH
Endemic to Madagascar. Non-exploited species.

DISTRIBUTION

Elevation: 0-1000 m a.s.l.
Vegetation formation: Forest
Bioclimate: Humid and sub-humid
Protected Area: Anjanaharibe-Sud, Mananara-Nord, Marojejy, Masoala and Zahamena
New protected areas: Makirovana, Mahabo-Mananivo an Analalava

Slides made from material collected in Mahabo-Mananivo.

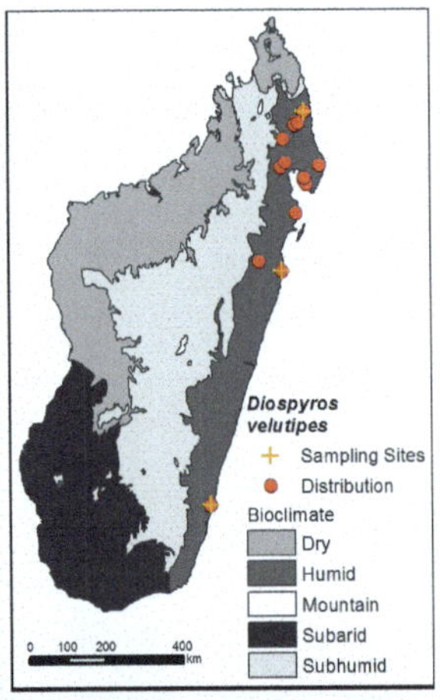

PITH AND PRIMARY XYLEM ANATOMY

Pith: Shape of pith round to oval. Thick-walled parenchyma cells present, predominating in number. Cells dimorphic. Cell content present in a few cells. Prismatic crystals frequent, often more than one crystal per cell. Pits in transverse cell walls. Pith visible in polarized light.

Primary xylem: Vascular bundles separated. Vessels solitary and in radial multiples of 3 to 5 elements.

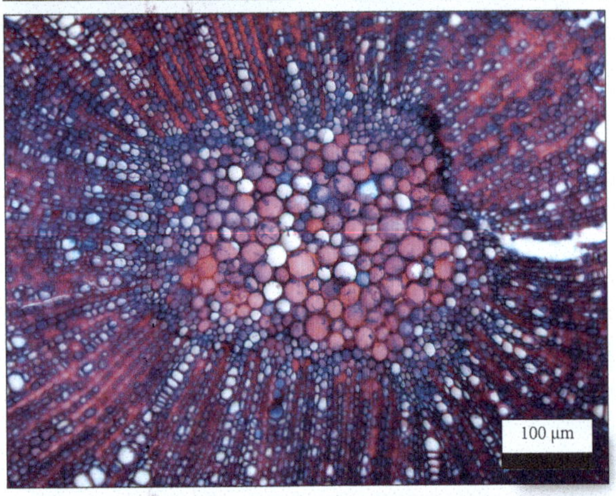

BRANCH XYLEM ANATOMY

Description: Growth rings boundaries indistinct or absent. Vessels mostly in radial multiples of 4 or more elements. Tangential diameter of vessels 50-65-100 µm, 39-64-75 vessels/mm². Vessel element length 180-301-410 µm. Intervessel pits small, 4-7 µm in diameter. Fibres thick-walled. Apotracheal parenchyma diffuse-in-aggregates, paratracheal parenchyma scanty. Body ray cell procumbent with over 4 rows of upright marginal cells. Rays exclusively uniseriate. Ray height 110-317-510 µm. Rays 11-18-22/mm. Prismatic crystals in chambered axial parenchyma cells.

Anatomical differences to stem xylem: Mean tangential diameter of vessel lumina is slightly smaller in branch (50-65-100/50-82-110 µm), vessel density is slightly higher in branch (39-64-75/42-54-72 vessels/mm²). Ray height is smaller in branch (110-317-510/140-371-750 µm), whereas number of rays per millimetre is higher (11-18-22/9-10-11 rays/mm).

BARK ANATOMY

Branch bark: Collapsed and non-collapsed sieve tubes present in the phloem. The cortex is characterized by a band of sclerenchyma and irregularly shaped blue-stained parenchyma cells. Dark staining substances in the phellem.

Stem bark: Sieve tubes and companion cells grouped. Ray dilatation slightly visible. Sclereids in few small irregularly distributed groups in the phloem. Radially arranged sclereids in nonconducting phloem. Several prismatic crystals in axial parenchyma cells. The lumen of some parenchyma cells in the phloem filled with dark staining substances. Phellem consists of regularly arranged rectangular cells, some filled with dark staining substances. Phellem indistinct in polarized light.

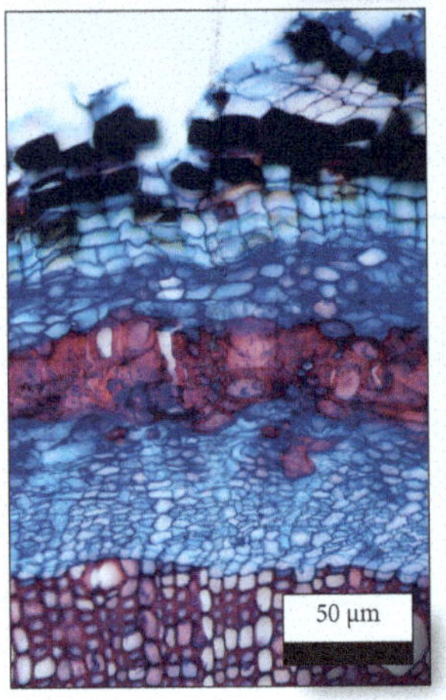

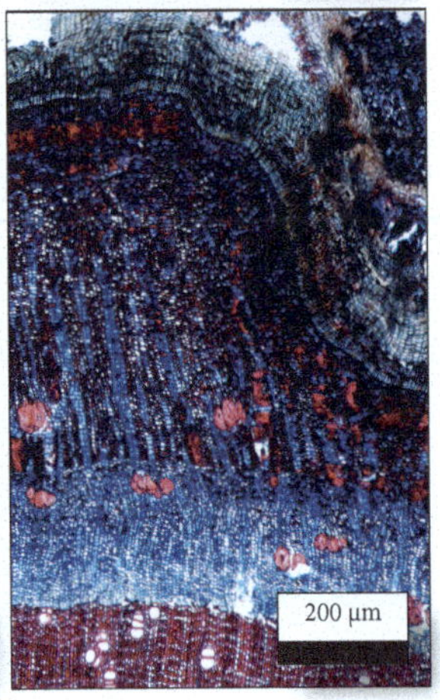

IAWA feature codes
2 5 7 9.1 10 13 22 25 30 41 49 53 58 61 66 69 70 77 78 79
86 92 93 96.1 102.3 107 115 136 142 180 189 192 196 202

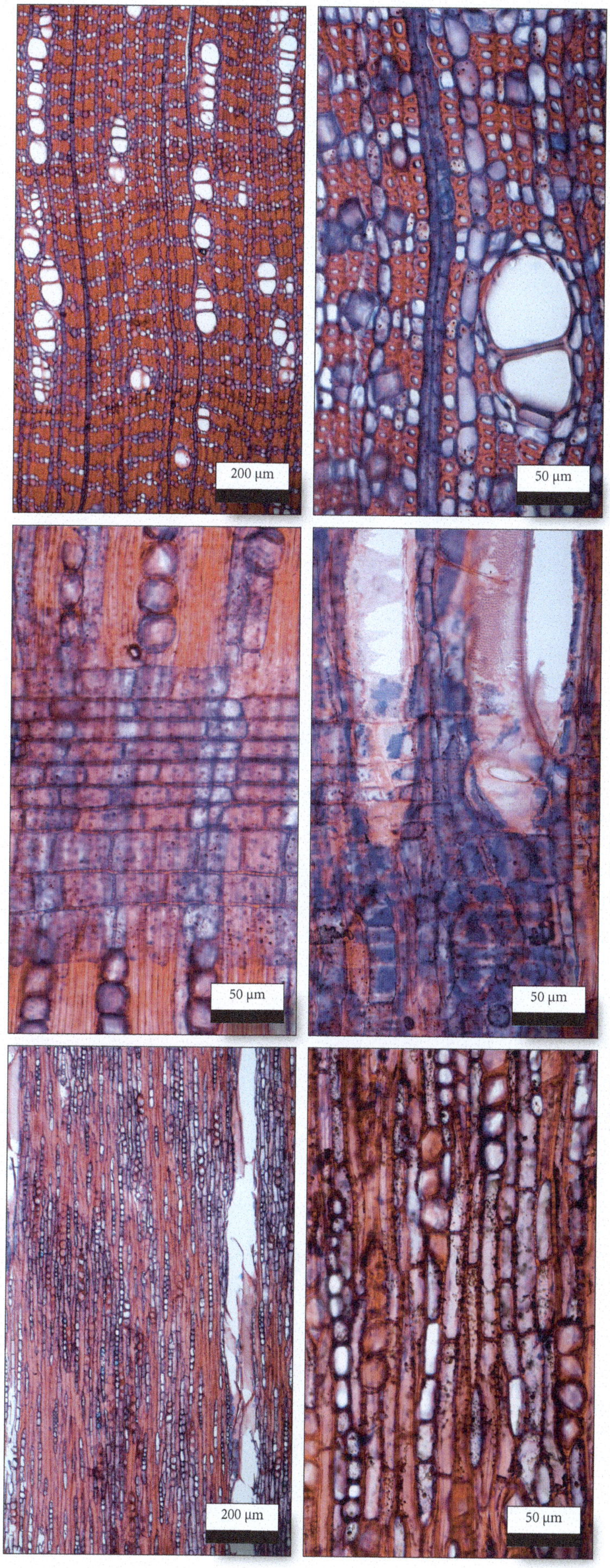

STEM XYLEM ANATOMY

Transverse section
Growth rings boundaries indistinct or absent.
Wood diffuse-porous.
Vessels in radial pattern.
Vessels solitary and in radial multiples 2 to 6(8) elements.
Tangential diameter of vessels medium 50-82-110 µm, 42-54-72 vessels/mm².
Gums present in heartwood vessel lumina.
Fibres thick- to very thick walled.
Apotracheal parenchyma diffuse-in-aggregates, and in narrow lines up to 3 cells wide. Paratracheal parenchyma scanty to vasicentric.
Rays 9-10-11/mm.

Radial section
Simple perforation plates.
Intervessel pits circular and alternate, 4-7 µm in diameter.
Vessel-ray pits with distinct borders; similar to intervessel pits in size and shape throughout the ray cell.
Vessel element length 210-389-540 µm.
Fibres with simple to minutely bordered pits (libriform fibers).
Heterogeneous rays, body ray cells procumbent with 2 to 4 rows of upright and square marginal cells.
Prismatic crystals in chamber axial parenchyma cells. Crystals very frequent.

Tangential section
Rays uni- and biseriate, predominantly uniseriate.
Ray height 140-371-750 µm.
Three to 8 cells per parenchyma strand.

Diospyros vescoi Hiern

Synonymy: *D. vescoi* var. *vescoi*

EN: Ebony FR: Bois d'ébène ML: Hazomainty

PLANT DESCRIPTION

Shrub or tree | 4-12 m high | 8-15 cm DBH
Endemic to Madagascar. Non-exploited species.

DISTRIBUTION

Elevation: 0-500 m a.s.l.
Vegetation formation: Forest
Bioclimate: Dry
Protected areas: Ankarana
New protected areas: Daraina complex, Montagne des Français, Orangea/Rigny and Sahafary

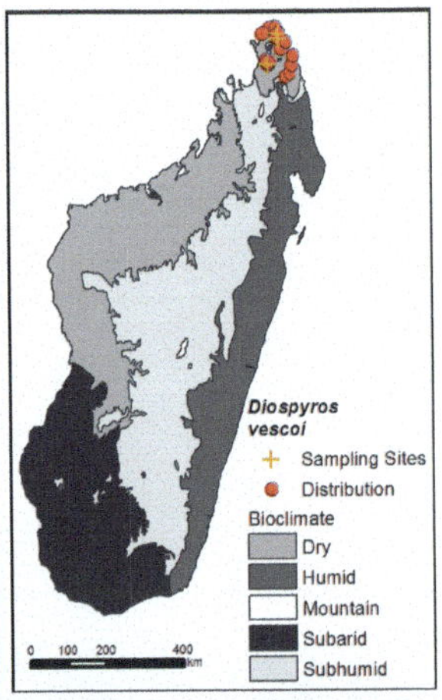

Slides made from material collected in Orangea/Rigny.

PITH AND PRIMARY XYLEM ANATOMY

Pith: Shape of pith oval to polygonal. Cells dimorphic. Most of cells are thick-walled. Cell content in a few cells. Prismatic crystals present. Pits in transverse cell walls. Pith visible in polarized light.

Primary xylem: Vascular bundles separated. Vessels mostly in radial multiples of 3-7 elements, only sometimes solitary.

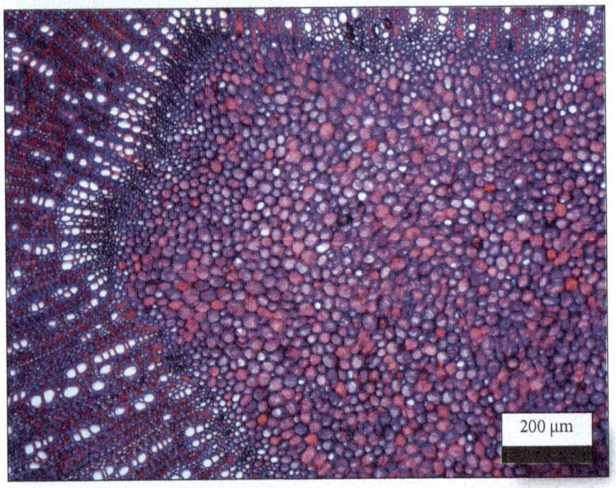

BRANCH XYLEM ANATOMY

Branch xylem: Growth rings boundaries indistinct or absent. Vessels solitary and in radial multiples of 2 to 5 elements. Intervessel pits minute, < 4 µm in diameter. Tangential diameter of vessels 20-46-60 µm, 55-71-85 vessels/mm². Vessel element length 160-278-380 µm. Gums and other deposits in some vessels. Fibres very thick-walled. Axial parenchyma diffuse-in-aggregates, and scanty paratracheal. Rays exclusively uniseriate. Ray height 160-511-1010 µm. Rays 26-32-37/mm. Three to 8 cells per parenchyma strand. Crystals very forming long axial lines in chambered axial parenchyma cells.

Anatomical differences to stem xylem: Mean tangential diameter of vessel lumina is slightly smaller in branch xylem (20-46-60/40-54-70 µm). Vessel density is higher in the stem (55-71-85/111-131-15 vessels/mm²). Rays are slightly higher in branch xylem (160-511-1010/140-474-920 µm), where ray number is also higher (26-32-37/17-18-20 rays/mm).

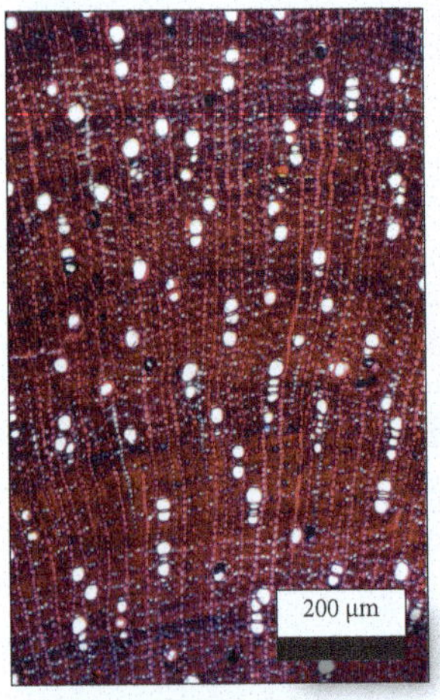

BARK ANATOMY

Branch bark: collapsed and non-collapsed sieve tubes present. Sclereid in irregularly distributed groups. The cortex is characterized by a band of fibers and with some irregularly dispersed sclereids and dimorphic parenchyma cells. The phellem is homogeneous.

Stem bark: sieve tubes in tangential rows. Collapsed sieve tubes present. Many groups of tangentially elongated sclereid groups. Prismatic crystals present within sclereid groups. Cell content in axial parenchyma cells in nonconducting phloem. The phellem is homogeneous and it consists of regularly arranged rectangular cells, indistinct in polarized light. Dark staining substances in outermost phellem cells.

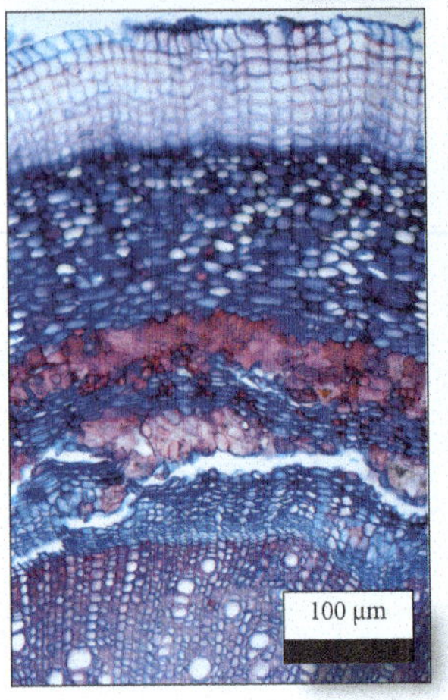

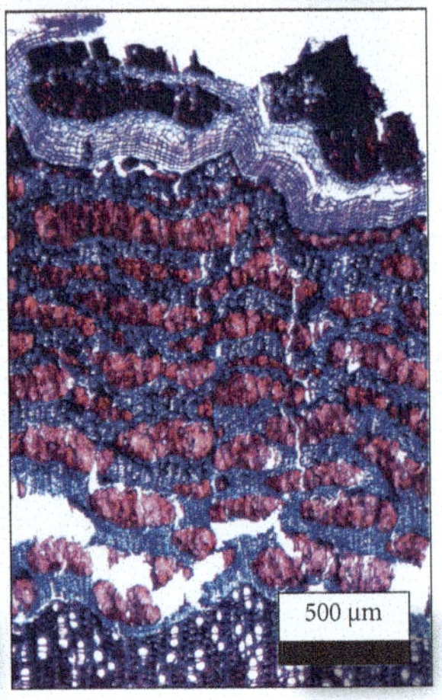

IAWA feature numberss
2 5 9.1 10 13 22 23 25 30 41 50 52 53 58 61 66 69 70 77 78 91 92 96.1 102.3 106 107 116 136 142 180 189 190 192 196 202

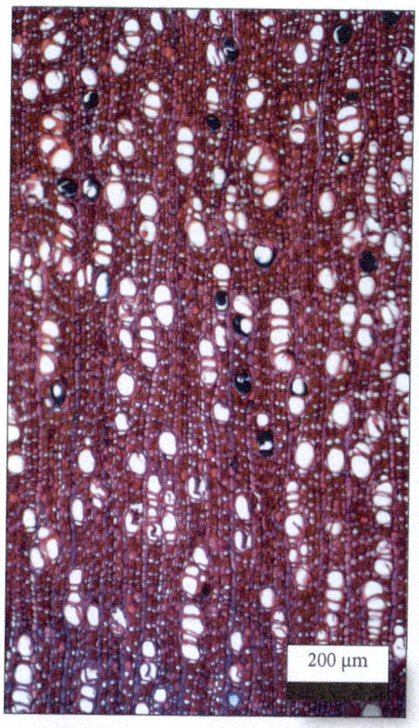

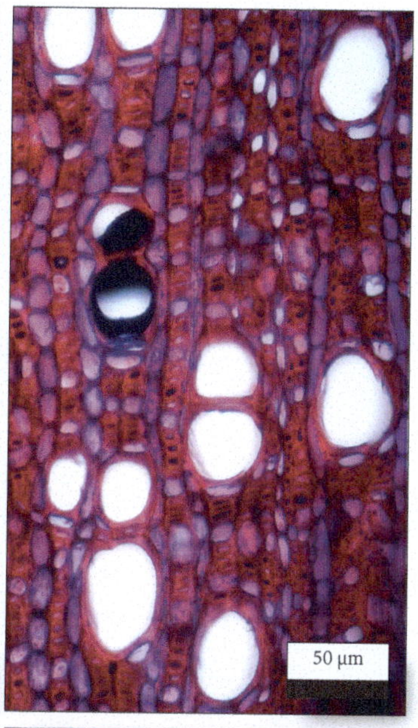

STEM XYLEM ANATOMY

Transverse section

Growth rings boundaries indistinct or absent.
Wood diffuse-porous.
Vessels solitary and in radial multiples of 2-9 elements.
Tangential diameter of vessels 40-54-70 µm, 111-131-152 vessels/mm^2.
Gums and/or other deposits in heartwood vessels.
Fibres thick- to very thick-walled.
Apotracheal parenchyma diffuse in aggregates.
Paratracheal parenchyma scanty.
Rays 17-18-20/mm.

Radial section

Simple perforation plates.
Intervessel pits circular and alternate, 4-7 µm in diameter.
Vessel-ray pits with distinct borders; similar to intervessel pits in size and shape throughout the ray cell.
Vessel element length 280-352-430 µm.
Fibres with simple to minutely bordered pits (libriform fibers).
Heterogeneous rays, body ray cells procumbent with 1 to 4 rows of upright and square marginal cells.
Prismatic crystals in chamber axial parenchyma cells. Crystals not frequent.

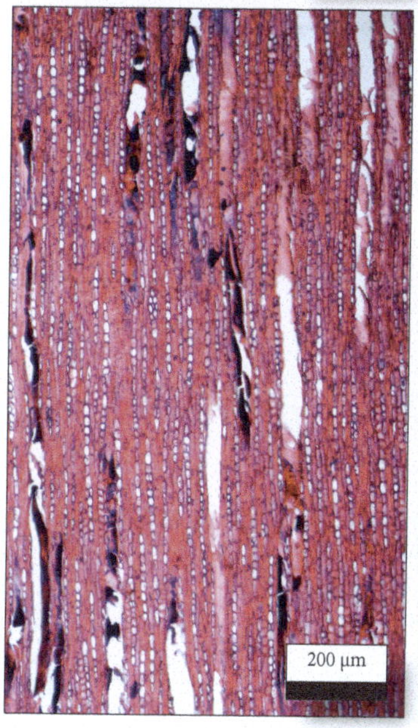

Tangential section

Rays predominantly uniseriate, a few biseriate rays present.
Ray height 140-474-920 µm.
Two to 4 cells per parenchyma strand.

References

Ballet J., Lopez P. & Rahaga N. 2010. L'exportation de bois précieux (*Dalbergia* et *Diospyros*) « illégaux » de Madagascar: 2009 et après. Madagascar Conservation Development. 5(2): 110-116.

Barett MA., Brown JL., Morikawa MK., Labat J-N. and Yoder AD. 2010. CITES designation for endangered rosewood in Madagascar. Science 328: 1109-1110.

Bosser J. & Rabevohitra R. (1996). Taxa et noms nouveaux dans le genre *Dalbergia* (Papilionaceae) à Madagascar et aux Comores. Bulletin du Muséum National d'Histoire Naturelle, 4 ème série, 18: 171-212.

Bosser J. & Rabevohitra R. (2005). Espèces nouvelles dans le genre *Dalbergia* (Fabaceae, Papilionaceae) à Madagascar. Adansonia 27, 2: 209-216.

Carlquist S. (1992). Wood, bark and pith anatomy of old world species of *Ephedra* and summary for the genus. Aliso. 13(2): 255-295.

Chaffey NB. (2002). Wood formation in trees. Cell and molecular biology techniques. Taylor and Francis. London, New York.

Crivellaro A. & Schweingruber FH. (2013). Atlas of Wood, Bark and Pith Anatomy of Eastern Mediterrranean Trees and Shrubs, with special focus on Cyprus. Springer Verlag. Berlin.

Department of Plant Biology and Ecology (2010). Evaluation écologique de quelques espèces de bois précieux les plus commercialisées à Madagascar (bois d'ébène, palissandre et bois de rose). Rapport final WWF.

Du Puy DJ., Labat J-N., Rabevohitra R., Villiers J-F., Bosser J. and Moat J. (2002). The Leguminoseae of Madagascar. The Royal Botanic Gardens Kew, U.K.

Espinoza EO., Wiemann MC., Barajas-Morales J., Chavarria GD. & McClure PJ. (2015). Forensic Analysis of Cites-Protected Dalbergia Timber from the Americas. IAWA Journal, 36(3), 311-325.

IAWA Committee. 1989. IAWA list of microscopic features for hardwood identification (E.A. Wheeler, P. Baas & PE. Gasson, eds). IAWA J. 10: 219-332.

IAWA Committee. 2004. IAWA list of microscopic features for softwood identification. (HG. Richter, D. Grosser, I. Heinz & PE. Gasson eds). IAWA J. 25: 1-70.

Junikka L. (1994). Survey of English macroscopic bark terminology. IAWA journal 15: 3-45.

Labat J-N. & Moat J. (2003). Leguminosae (Fabaceae). In: The Natural History of Madagascar, S.M. Goodman and J.P. Benstead (des), 340-373. University of Chicago Press, Chicago, USA.

Lancaster C. & Espinoza, E. (2012). Analysis of select Dalbergia and trade timber using direct analysis in real time and time-of-flight mass spectrometry for CITES enforcement. Rapid Communications in Mass Spectrometry, 26(9), 1147-1156.

McClure PJ., Chavarria GD. & Espinoza, E. (2015). Metabolic chemotypes of CITES protected Dalbergia timbers from Africa, Madagascar, and Asia. Rapid Communications in Mass Spectrometry, 29(9), 783-788.

Metcalfe CR & Chalk L. (1950). Anatomy of the Dycotyledons. Oxford University Press, London.

Missouri Botanical Garden. (2009). L'analyse écologique et économique des bois importants dans le complexe forestier de Makirovana- Tsihomanaomby dans la Région de SAVA. Rapport de mission.

Missouri Botanical Garden (2014). Capitalisation des échantillons de *Dalbergia* L. f (Fabaceae) et de *Diospyros* L. (Ebenaceae) de Madagascar dans l'Annexe II de la CITES. Rapport final DBEV.

Parsa Pajouh D. & Schweingruber F. H. (1988). Atlas des bois du Nord de l'Iran: Teheran University publications.

Patel ER. (2007). Logging of rare rosewood palissandre (*Dalbergia* spp.) within Marojejy National Park. Madagascar Conservation & Development 2.1: 11-16.

Perrier de la Bâthie H. (1952). 165ème Famille-EBENACEAE. In: Flore de Madagascar et des Comores. Typographie Firmin-Didot et Cie, Paris.

Schatz EG & Lowry II PP. (2011). Nomenclatural notes on Malagasy *Diospyros* L. (Ebenaceae). Adansonia, sér 3, 33 (2): 271-281.

Schweingruber FH. (2007). Wood structure and environment. Springer Verlag. Berlin.

Schweingruber FH. (1990). Anatomy of European woods: Paul Haupt. Bern.

Trockenbrodt M. (1990). Survey on discussion of the terminology used in bark anatomy. IAWA Bull. 11: 141-166.

Wheeler E.A., Baas P. & Gasson P.E. (eds.) (1989). IAWA List of microscopic features for hardwood identification. IAWA Bull. 10(3): 219-332.

B.H. Ravaomanalina et al., *Stem Anatomy of Dalbergia and Diospyros Species from Madagascar*, DOI 10.1007/978-3-319-51147-4

Index of Species and Families

Names in *italic* refer to botanical names, names in normal type are synonyms, names in **bold** are families.

B.H. Ravaomanalina et al., *Stem Anatomy of Dalbergia and Diospyros Species from Madagascar*, DOI 10.1007/978-3-319-51147-4

The manufacturer's authorised representative in the EU is Springer Nature Customer Service Centre GmbH, Europaplatz 3, 69115 Heidelberg, Germany. If you have any concerns regarding our products, please contact ProductSafety@springernature.com

Printed and bound by CPI Group (UK) Ltd, Croydon, CR0 4YY
05/10/2026
02228165-0001